FOCUS ON PHYSICS

The Barnes & Noble Focus on Physics titles are prepared under the general editorship of J. WARREN BLAKER, Associate Professor of Physics, Vassar College.

ABOUT THE AUTHOR

J. Warren Blaker received his B.S. degree from Wilkes College and his Ph.D. degree from the Massachusetts Institute of Technology. He has held a number of industrial positions and taught at Fairleigh Dickinson University and at Vassar College, where he is now Associate Professor of Physics.

Professor Blaker is a member of the American Association for the Advancement of Science, the American Physical Society, the Physical Society (London), and the American Association of Physics Teachers. He is co-editor of *Perspectives in Modern Physics*.

College Outline Series

FOCUS ON PHYSICS
Optics II—Physical and Quantum Optics

J. WARREN BLAKER

Associate Professor of Physics
Vassar College

BARNES & NOBLE, INC., NEW YORK
Publishers   *Booksellers*   *Since 1873*

L. C. catalogue card number: 74-80003
SBN   389   00088   4

Printed in the United States of America

Preface

FOCUS ON PHYSICS

The ten Focus on Physics volumes in the College Outline Series are concise but comprehensive self-teaching treatments of the most important topics in the first-year physics course.

It has been found that beginning students of physics often require material which can supplement their texts and lectures by supplying, in a somewhat different format, explanations of principles and methods. Such re-enforcement is particularly important in areas where the student is having difficulty. A number of short "outline" books which emphasize the problem-solving aspect of physics are available for this purpose, but we know of no short works which provide a more thorough discussion of underlying principles.

Each Focus on Physics title has been planned to present the subject matter of a particular topic or group of related topics. Each book includes summaries of principles and facts, solved examples, problems with answers, and numerous illustrations.

In order to help the student to attain understanding and mastery of the basic material, these volumes strongly emphasize physical principles. They reflect the detailed approach to physics which is now the standard treatment, and their combined subject matter forms the essence of physical science.

In addition to use by the individual student, the Focus on Physics titles are suitable for class assignment as text supplements. They can also be used for self-instruction by the general reader who is interested in exploring and learning about the elements of modern physics.

J. WARREN BLAKER

General Editor

Table of Contents

Introduction

OPTICS II—Physical and Quantum Optics

This text is a self-contained treatment of wave and quantum optics at a level suitable for beginning students. The wave and particle models are two of the three important models for the treatment of optical phenomena. The other model, the ray model, is introduced in *Optics I* of this series.

We begin with a general treatment of waves and proceed by relating the basic physical optical phenomena: interference, diffraction, and polarization to the wave picture. Finally, we introduce Planck's hypothesis and examine the basic set of experimental results which lead to the photon concept and to quantum optics. The laser is treated in the final chapter as an example of the application of the concepts developed throughout the book.

Chapter 1

Waves

Introduction. It was pointed out in *Optics I*[1] that a number of models for the propagation of light are commonly used by physicists. These include: the ray approximation, where one deals with light by studying the trajectory of the light as it passes through various media; physical optics, or wave optics, where the light is treated on the basis of its wave properties; and quantum optics where light is treated on the basis of individual quanta, or photons. Each of these approximations has a set of phenomena which are best understood in that particular approximation. For example, the interference of light can be easily treated on the basis of the wave properties of light, but it cannot be treated simply from the standpoint of either of the other two approximations.

This chapter will treat the various properties of waves and will describe the propagation of waves and the phenomena arising as a result of a wave approximation. In the subsequent chapters, these concepts will be applied to light.

The Form of Waves. Everyone is familiar with the waves which occur on the surface of lakes. These waves repeat in their form over a short distance known as the *wavelength* (see Figure 1.1). We will use the symbol λ to represent the wavelength. This

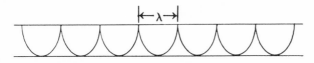

Figure 1.1. Water waves showing the wavelength, λ.

wavelength will be essentially constant throughout the medium in which the propagation occurs if the properties of the medium remain constant.

Water waves on the surface of a lake are traveling waves. That is to say that one can follow one peak of the wave as it moves

[1] Blaker, J. Warren, *Optics I*, Focus on Physics Series, Barnes & Noble, New York.

along the surface of the water. The velocity at which the wave moves is the wave velocity, v. The *period* of the wave, T, is the time required for the wave to pass through one full cycle—from peak to depression to peak—at a fixed point in space. We can view the period as the time required to shift the entire train of waves one wavelength in the direction of propagation. We can then write:

$$T = \frac{\lambda}{v} \tag{1.1}$$

The reciprocal of T is the *frequency*, the number of waves passing a given point in space in unit time:

$$\nu = \frac{1}{T}, \tag{1.2}$$

where ν is the frequency. By combining equations 1.1 and 1.2, we get the fundamental equation of wave propagation:

$$v = \nu\lambda \tag{1.3}$$

This equation tells us how fast the wave propagates through the medium. This is an important parameter of the wave since the wave is actually a manifestation of energy transport. What the wave represents is the energy, which has been imparted to the medium by some external agency such as the wind, moving through the medium.

Example 1.1. A fisherman sitting in an anchored boat observes the waves passing his boat. He notes that the peak-to-peak distance is 4 ft and that the waves are moving past his boat at a velocity of 16 ft/min. What is the frequency and the period of the waves?

From equation 1.3 we immediately see:

$$16 \text{ ft/min} = \nu \times 4 \text{ ft}$$

and $\nu = 4 \text{ min}^{-1}$. The period is given by equation 1.2 as:

$$T = \frac{1}{\nu} = \frac{1}{4} = 0.25 \text{ min}$$

Equation 1.3 is valid for waves in general although it must be modified on occasion because the velocity is often frequency dependent.

We will not deal with waves having the form of Figure 1.1 but rather with waves having the form of sine curves, as illustrated in

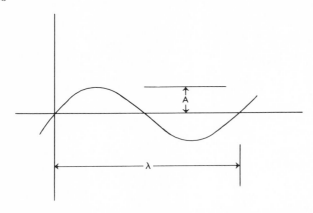

Figure 1.2. Amplitude, A, and wavelength, λ, of a harmonic (sine) wave.

Figure 1.2. There is an important reason for treating waves with sinusoidal dependence. A theorem due to Fourier states that any periodic phenomenon may be represented by an infinite series of sinusoidal terms.[1] We will only deal with sinusoidal waves since any more complex waveform can be reduced to a sinusoidal form.

With such traveling waves it is important to note that it is the disturbance and not the medium which propagates. A stick on the surface of the lake will not be carried along with the waves at the wave velocity, but will instead bob up and down at the same position. This can be understood in terms of the idea of waves as a mechanism for energy transport. The energy in the medium re-

[1] The Fourier series expansion takes the form:

$$f(x) = \sum_{n=0}^{n=\infty} (A_n \sin n \cdot x + B_n \cos n \cdot x)$$

$$= B_0 + A_1 \sin x + A_2 \sin 2x + A_3 \sin 3x + \cdots$$
$$+ B_1 \cos x + B_2 \cos 2x + B_3 \cos 3x + \cdots$$

where the A_n and B_n are adjusted so that the series represents the required periodic situation.

For example, the square wave in Figure 1.3 (p. 4) can be represented by a series:

$$g(x) = \frac{4}{\pi} \sum_{n=\text{odd}}^{n=\infty} \frac{1}{n} \sin nx$$

where the sum is over the odd values of n out to infinity. In practice only as many terms are taken as can be conveniently handled, but in general about ten terms will give a moderately good fit to the curve. This series is less general than the full Fourier series since the function $g(x)$ has the symmetry of the sine curve itself.

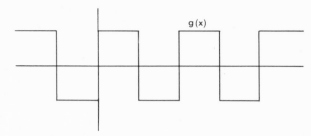

Figure 1.3. A square wave.

sults in the displacement of the medium, giving rise to the wave. When the medium is distorted, energy is stored in the medium, and the propagation of the stored energy gives rise to the apparent motion of the wave.

We can explore this in more detail. Consider a pulse disturbance as illustrated in Figure 1.4. As this disturbance moves through the medium, the leading edge forces the medium into motion and causes the medium to distort. The maximum distortion is known as the amplitude, A, of the wave (as illustrated in Figure 1.2), and *the square of the amplitude is proportional to the energy in the wave.* (This is discussed elsewhere in this series.[1]) At the maximum displacement, all the energy is stored as potential

Figure 1.4. A propagating pulse wave.

energy; while during the rise and fall of the disturbance some of the energy is in the form of kinetic energy of the medium, but the motion associated with this kinetic energy is normal to the direction of propagation. This is illustrated with a double arrow in Figure 1.4. Waves which cause a displacement of the medium normal to the direction of propagation are called *transverse waves.* Waves such as sound waves with the displacement of the medium parallel to the direction of propagation are called *longitudinal waves.* In this book we will deal only with transverse waves.

Equation 1.3 describes the propagation of the wave, but it does

[1] *Mechanics II*, Focus on Physics Series, Barnes & Noble, New York.

not fully describe the wave. We get no information about the behavior of a specific point in the medium and we would like to know the displacement of each point in the medium and its behavior as a function of time. What is required is a function of two variables, time and position, which will then give us the total behavior of the wave. Such a function will obviously be sinusoidal because of the constraints we have established. This function will describe the *wave profile* as a function of time.

Figure 1.5 shows a wave in an initial position and after a displacement a short time later. The solid line we will take as the

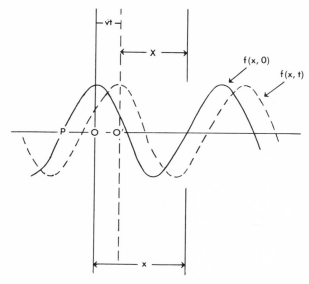

Figure 1.5. A harmonic wave at t = 0 and at a later time, t.

t = 0 position of the wave. The wave is then described by the equation for f(x, t) at t = 0:

$$f(x, 0) = A \cos kx \qquad (1.4)$$

A short time later the wave has moved to the position shown in Figure 1.5 by the dotted lines. The wave profile has just moved a distance vt in the figure where v is the wave velocity. If we take a new origin at O′ and measure a distance **X** from this origin it would be related to the original origin by the expression:

$$x = X + vt \qquad (1.5)$$

The equation of the wave referred to the new origin is:

$$f(X) = A \cos kX$$

We can substitute equation 1.5 for X to get the wave profile referred to the original origin:

$$f(x, t) = A \cos k(x - vt) \qquad (1.6a)$$

which is the equation we wanted. We can now describe the entire wave profile as a function of time and position.

For a wave traveling to the left, corresponding to a negative velocity, equation 1.6 is modified by changing the sign of the velocity and we get:

$$f(x, t) = A \cos k(x + vt) \qquad (1.6b)$$

Waves such as are described by equations 1.6a and 1.6b are known as harmonic, or sinusoidal, waves. We can change the cosine function to a sine function merely by changing the origin to a point such as P in Figure 1.5. As a matter of fact, we are free to place the origin anywhere, and an arbitrary placement of the origin will only require that we add a phase factor such as δ to the argument of the function. In such a case equation 1.6 would read:

$$f(x, t) = A \cos k(x - vt + \delta)$$

It remains for us to evaluate the constant k which was arbitrarily introduced into equation 1.4. The wave repeats at a regular distance of $2\pi/k$ as can easily be seen from the figure and through equation 1.4. This distance is just the wavelength which we have previously defined, and then equation 1.6 becomes:

$$f(x, t) = A \cos 2\pi \left(\frac{x}{\lambda} \pm \frac{vt}{\lambda} \right) \qquad (1.7a)$$

where the meaning of each term is well established. Equation 1.7a can take a number of forms and the most important of these arises by introducing the frequency for the v/λ term:

$$f(x, t) = A \cos 2\pi \left(\frac{x}{\lambda} \pm \nu t \right) \qquad (1.7b)$$

Superposition. The solution for the wave profile which we have just found from purely physical arguments can also be derived analytically from the wave equation. This equation[1] is a partial differential equation for the wave and is derived by considering the motion of a single point in the medium as a function of time.

There is one very important result derived from the analytic solution of the wave equation. The wave equation is linear, that is to say if $g(x, t)$ and $h(x, t)$ are two different solutions to the wave equation their sum $\epsilon_1 g(x, t) + \epsilon_2 h(x, t)$ is also a solution, where the ϵ's are arbitrary constants fixed generally by the relative amplitudes. This is an example of the principle of superposition which states that the disturbance in a medium is the sum of all the individual disturbances; no one disturbance is changed by the presence of the others. Figure 1.6 illustrates this principle. Here $g(x, t)$ and $h(x, t)$ show two disturbances and their sum is also shown. Superposition may be thought of as a kind of inverse Fourier Theorem where a wave of any given form may be generated by combining harmonic waves.

As an example of superposition we can consider the effect of combining two harmonic waves with different phases. Their sum can be given as:

$$A \sin k(x - vt) + A_2 \sin \{k(x - vt) + \alpha_1\} \qquad (1.8)$$

where we have chosen the phase of the first wave as zero. This is possible because we are free to choose the origin at any point, but since the waves have different phases only one phase can be zero. We can expand the second term in equation 1.8 and we get:

$$A_1 \sin k(x - vt) + A_2 \sin k(x - vt) \cos \alpha_1$$
$$+ A_2 \cos k(x - vt) \sin \alpha_1$$

using the standard trigonometric expression for the sine of the sum of angles. Collecting terms we get:

$$(A_1 + A_2 \cos \alpha_1) \sin k(x - vt) + A_2 \sin \alpha_1 \cos k(x - vt)$$

---

[1] The wave equation is written:

$$\frac{\partial^2 f(x, t)}{\partial x^2} = \frac{1}{v^2}\left(\frac{\partial^2 f(x, t)}{\partial t^2}\right)$$

and is given here only for completeness.

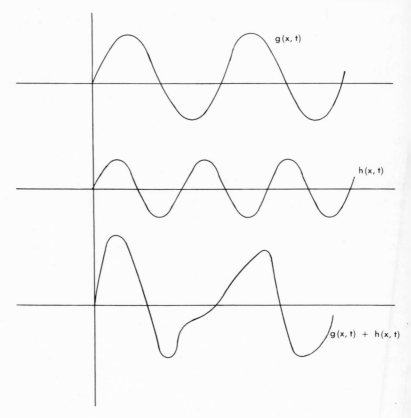

Figure 1.6. Superposition of two harmonic waves.

which can be rearranged to read:

$$(A_1 + A_2 \cos \alpha_1) \left\{ \sin k(x - vt) + \cos k(x - vt) \left(\frac{A_2 \sin \alpha_1}{A_1 + A_2 \cos \alpha_1} \right) \right\}$$

The coefficient of the cosine term in the braces is a constant since it depends only on the relative amplitudes and the phase difference. We can then replace this term by another constant which we will call $\tan \delta$. If we break $\tan \delta$ into the ratio $\frac{\sin \delta}{\cos \delta}$, substitute back into the equation, and rearrange we get:

$$\left(\frac{(A_1 + A_2 \cos \alpha_1)}{\cos \delta} \right) \{ \sin k(x - vt) \cos \delta + \cos k(x - vt) \sin \delta \}$$

$$(1.9)$$

which is just another sine wave of the form:

$$A \sin \{k(x - vt) + \delta\} \tag{1.10}$$

and we have the general result that the combination of two harmonic waves differing only in phase and not in period leads to a new wave of the same frequency, but with altered amplitude and phase.

Figure 1.7 illustrates the addition of two harmonic waves of the same frequency. In Figure 1.7a, the phase difference is π radians or 180°, and just a modification of the amplitude occurs with no relative phase shift. In Figure 1.7b, where the phase shift is of the order of $\pi/2$ radians or 90° both the amplitude and the phase are shifted.

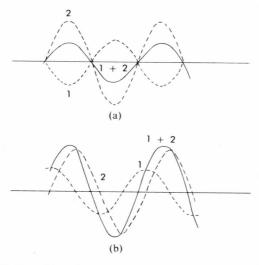

(a)

(b)

Figure 1.7. Superposition of two waves of the same period but different phases.

We can find the amplitude and phase in Figure 1.7b by using equation 1.9. The amplitude $A_1 = 1$ and $A_2 = 2$ and the phase difference α_1 is $-90°$. We first evaluate $\tan \delta$ from its definition given above and find:

$$\tan \delta = \frac{A_2 \sin \alpha_1}{A_1 + A_2 \cos \alpha_1}$$

$$= -2$$

and $\delta = -63.4°$.

The new amplitude is given by:

$$A = \frac{A_1 + A_2 \cos \alpha_1}{\cos \delta}$$

$$= \frac{1 + 2(0)}{0.447} = 2.24.$$

These results agree quite nicely with the figure. Note that the phase shift of curve 2 in the figure is negative since it lags behind curve 1.

Standing Waves. One of the interesting results of the super-position principle is the standing wave. Suppose that we have two waves of the same amplitude and period traveling in opposite directions. In terms of energy transport ideas this is equivalent to no net transmission of energy.

The resultant wave is just the sum of the two waves in the medium by superposition and we can write for the total disturbance:

$$A \sin k(\text{x} - \text{vt}) + A \sin k(\text{x} + \text{vt}) \tag{1.11}$$

This expression can be expanded by the trigonometric identities for the sine of the sum of two angles and we get:

$$A \sin k\text{x} \cos k\text{vt} - A \sin k\text{vt} \cos k\text{x} + A \sin k\text{x} \cos k\text{vt}$$
$$+ A \sin k\text{vt} \cos k\text{x}$$

which when combined takes the form:

$$2A \sin k\text{x} \cos k\text{vt}$$

which is the total disturbance occuring in the medium. The first part of the expression ($2A \sin k\text{x}$) is dependent only on the position in the medium and is an amplitude term. The amplitude of the total disturbance depends on the position and at any position the medium oscillates as $\cos k\text{vt}$.

Figure 1.8 shows a standing wave. The dotted lines are the amplitude of the wave and the solid lines represent the wave at some fixed instant in time. The important effect here is the com-

Figure 1.8. A standing wave.

plete cancellation of the disturbance at certain points in the medium due to destructive interference of the wave motions. At the maximum points of the medium's displacement there is constructive interference of the waves, and the excursion of the medium is twice that which would be encountered if only one of the waves were present.

Much of the remainder of this book will be devoted to the consideration of the effects presented in this chapter as applied to light waves. The importance of superposition will become quite clear as we continue.

Chapter 2

Interference

In the previous chapter we considered the principle of the superposition of waves for a general case. We will now apply this principle to light waves. We are going to assume that light is a wave disturbance which can travel throughout space. This disturbance does not require a medium for support, but is of such a nature that it can travel through both vacuum (as a sunlight beam) or through transparent media (such as glass). We will also investigate the effect of such media on the waves of light in both this and later chapters.

Young's Experiment. The experiment of Thomas Young (1773–1829), or the two-slit interference experiment, was one of the earliest experiments which was directly concerned with the wave nature of light. The experimental arrangement for this experiment, first performed in 1800, is shown in Figure 2.1a. Light from a distant source falls on slit S_1 which then acts as a secondary source for a pair of slits in a second screen. These slits are very narrow and are separated by only a fraction of a millimeter. The light passing through these slits, S_2 in the figure, then passes on to a viewing screen, VS. An enlarged portion of the image formed on the viewing screen is shown in Figure 2.1b.

The image consists of a number of equally spaced bright and dark lines known as *fringes*. We know from Huygen's principle[1] that the slits S_2 can act as secondary sources for the source at S_1. The waves from S_2 which reach the viewing screen in phase interfere constructively to give a bright fringe. Those which reach the screen with a 180° phase difference between the waves will interfere destructively, with the result that the light in those regions will be completely cancelled. Figure 2.2 shows the intensity distribution along the viewing screen. Along the centerline we have a fringe of maximum intensity. Immediately adjacent to this are dark fringes. Successive bright and dark fringes are separated by equal distances.

If the light leaving each of the slits S_2 is in phase, then a dark

[1] For a discussion of Huygen's principle see *Optics I*.

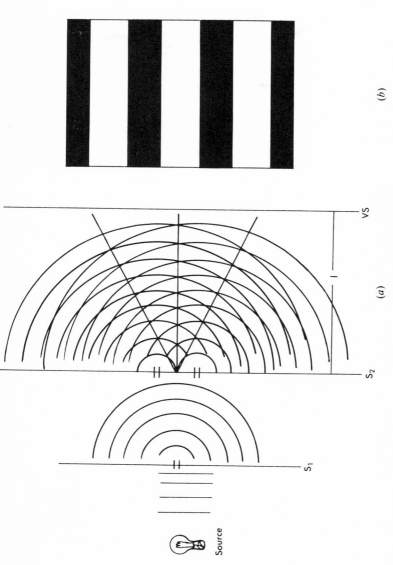

Figure 2.1. (a) The experimental set-up for Young's experiment showing the slits and the waves propagating to the viewing screen. (b) An enlarged portion of the viewing screen.

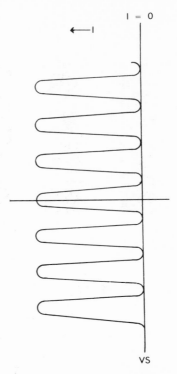

Figure 2.2. The intensity distribution of the light falling on the viewing screen.

fringe will occur each time the wave from one slit must travel one-half wavelength farther than the light from the other slit. On the centerline the light will be a maximum since the centerline is always equidistant from the two slits. We can find the position of the first dark fringe by finding the position at which the distance from one slit is one-half wavelength greater than from the other. Figure 2.3 shows the various parameters involved. The distance between the slits and the viewing screen is very much greater than the separation of the slits, and the slits themselves are so small that they can be considered as point sources. We first construct $S_{21}A$ so that $S_{21}D$ is equal to AD. The line from the centerline at O to D is then perpendicular to $S_{21}A$, and the slits determine a line normal to the centerline. As a result, the angles labeled α are equal. The distance $S_{22}A$ must be equal to a half-wavelength, $\lambda/2$, to achieve the proper phase shift. Now by equating

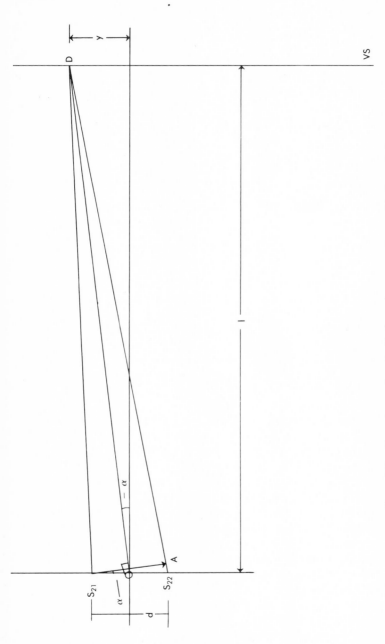

Figure 2.3. Young's experiment with the various parameters used in the analysis.

the tangents of the two angles α we get:

$$\frac{y}{l} = \frac{\lambda/2}{d} \qquad (2.1)$$

and this expression relates the wavelength of the light to other measurable parameters of the experiment.

We can use the results of Young's experiment to determine the wavelength of light. The experiment was performed with a slit separation of 0.1 mm and with the viewing screen at a distance of 50 cm from the slits. The first dark line fell at a distance of 1 mm from the central bright fringe. If these numbers are substituted into equation 2.1 (with all units in millimeters) we get:

$$\frac{1}{500} = \frac{\lambda/2}{0.1}$$

and:

$$\lambda = 4 \times 10^{-4}\,mm = 400\,m\mu$$

(The wavelength of light is usually given in millimicrons or Angstrom units: 1 millimicron (mμ) = 10^{-9} meters; 1 Angstrom unit (Å) = 10^{-10} meters.) The above is an important result in that it tells the order of the wavelengths of light. Visible light has wavelengths ranging from 400 mμ in the violet region of the spectrum to 700 mμ in the red end of the spectrum. Appendix 1 lists some of the common spectral lines and their wavelengths and color.

Young's experiment is not easy to perform since a number of very important conditions must be met. We assumed that the light was in phase as it passed through slits S_2. Most light sources are extended sources, that is they are large in physical dimension and they emit light via a heated filament or a region of heated gas. In general, in such sources there is no correlation between the phase of light emitted at various points in the source. Thus, we generally do not see interference effects because of the continuing and rapidly varying phase relationships. This is overcome in the experiment we have just described by use of S_1 which selected the light from a small portion of the original source. Light with fixed phase is termed *coherent* and a general light source such as a bulb is termed *incoherent*.

The experimental setup also assured us that the light from each of the slits S_2 would have about the same amplitude. Only with light of about equal amplitude can interference be seen. Also a

monochromatic source, a source which emits only one wavelength
in this case 400 mμ), is essential if the difference in path lengths
s larger than a few wavelengths.

The position of any fringe, bright or dark, is essentially found
by using the correct multiple of wavelengths for the distance $S_{22}A$.
Bright fringes will occur whenever the distance $S_{22}A$ is an even
multiple of wavelengths: 1λ, 2λ, 3λ, etc. Dark fringes will occur
whenever this interval is an odd number of half wavelengths:
$\frac{1}{2}\lambda, \frac{3}{2}\lambda, \frac{5}{2}\lambda$, etc. Some examples will help to illustrate the general
method of calculation.

Example 2.1. Yellow light of wavelength 589 mμ is used to
produce interference fringes in a Young's experiment. The slits
are separated by 0.2 mm, and the viewing screen is 1 m from the
slits. What is the position of the third bright fringe relative to
the central maximum?

We will want to apply equation 2.1 to this problem. Equation
2.1 reads:

$$\frac{y}{l} = \frac{\lambda/2}{d}$$

where y is our unknown. In place of the $\lambda/2$ we use 3λ, which
is the difference in path lengths for the two rays at the third
maximum. Our equation reads then:

$$\frac{y}{1000} = \frac{589 \times 3}{2 \times 10^5}$$

where we have substituted the appropriate numbers. All the numbers (and y) on the left-hand side of the equation are in millimeters and those on the right-hand side are in millimicrons, so
that the units are balanced. Solving, we get y = 8.84 mm. This
is the distance at which we find the third maximum displaced
from the central maximum. There are two third maxima: one
above the central maximum and one below.

Example 2.2. A Young's experiment is performed using violet
light. The slit separation is 0.1 mm (10^5 mμ) and the viewing
screen is located 75 cm from the slits. Find the wavelength of the
light if the first maximum is 5.25 mm from the central maximum.

The equation will read:

$$\frac{y}{l} = \frac{\lambda}{d}$$

and, with the appropriate quantities substituted, we have:

$$\frac{5.25}{750} = \frac{\lambda}{10^5}$$

and $\lambda = 700$ mμ. This result is consistent with what we know the wavelength of red light to be.

Modified Young's Experiments. A number of other methods for producing interference effects are similar in nature to Young's experiment, and can be analyzed by the same procedures. These modified experiments are illustrated in Figure 2.4. In each of these the object is to provide two sources, either real or virtual derived from a single source in order to provide coherence.

Figure 2.4a illustrates the Fresnel double mirror. The mirror consists of two plain mirrors set at a small angle to each other. These mirrors each form a virtual image (S' and S'' in the figure) of the source located at point S. These two sources then function in exactly the same way as the double slits in Young's experiment. The angle between the two mirrors must be very close to 180° so that the two sources S' and S'' are close together so as to meet the requirements of the experiment.

Fresnel also developed the biprism illustrated in Figure 2.4b. The two prism portions of the biprism function as thin optical wedges and give rise to the virtual images S' and S''. Once again the analysis is similar to that of Young's experiment. One merely has to evaluate the difference in distance between S' and S'' and the viewing screen VS. This difference in distance is then transformed into a relative phase shift from which the intensity of the image at a particular point on the screen can be evaluated.

Figure 2.4c illustrates Lloyd's mirror. Lloyd noted that a single mirror was sufficient to produce two sources. The first is the real source due to the source slit itself; the second is the virtual source produced by the mirror. In performing the Fresnel double mirror experiment, a short curtain C (see Figure 2.4a) is incorporated into the apparatus so that no waves reach the screen from the real source S. Lloyd's mirror makes use of both the real source S and its virtual image formed in the mirror.

Example 2.3. Evaluate the Lloyd's mirror experiment for a source one meter from the viewing screen with a plane mirror located midway between the source and the screen, 0.5 mm below the source. The details of the experiment are given in Figure 2.5.

We can locate S' immediately from the theory of plane mirrors. The virtual image will be formed behind the plane mirror at a

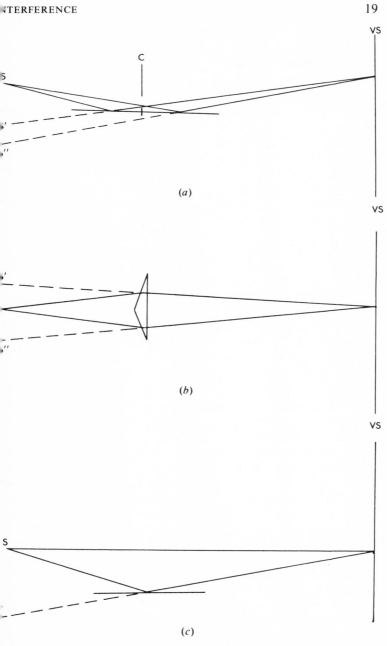

Figure 2.4. (a) Fresnel's double mirror. (b) Fresnel's biprism. (c) Lloyd's mirror.

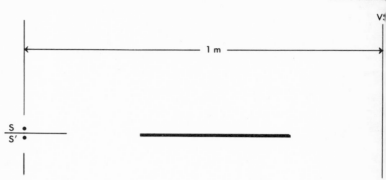

Figure 2.5. Lloyd's mirror with the virtual second source, S'.

distance equal to the distance from the plane of the plane mirror to the source. The extension of the plane of the plane mirror is shown in the figure. Since the distance from the plane of the mirror is 0.5 mm (from the problem statement), the separation of the two sources will be 1 mm. The remainder of the analysis is precisely the same as for the double slit experiment, except for the fact that no image is formed below the plane of the mirror since this area is "shadowed" by the mirror. The result of the experiment is just half of the pattern produced by Young's apparatus.

Wavelength Measurement—Michelson Interferometer. We have seen in the previous section that one of the uses of interferometry is the measurement of wavelengths. Such measurements are of particularly great importance for spectroscopy, where the wavelength measurements must often be made to six or more significant figures. The experiments we have previously considered, while perfectly capable of being used as wavelength measuring experiments, are not best suited for that purpose. Wavelength measurements are best made by means of an instrument known as an interferometer. Interferometers are available in a number of different modifications. In this section we will consider one such instrument, the Michelson Interferometer. Other modifications will be discussed later in the chapter.

The Michelson Interferometer is shown schematically in Figure 2.6. Light from the source S (usually monochromatic) is directed onto lens L which serves to render the beam entering the interferometer parallel. Mirror M_1 which is placed at 45° to the

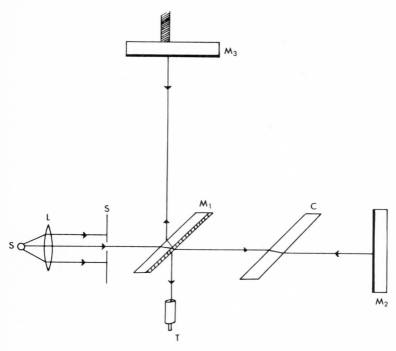

Figure 2.6. The Michelson interferometer.

incoming beam is a half-silvered mirror.[1] The mirror M_1 is silvered on the backside and serves as a beamsplitter. The re-flected portion of the beam passes back through M_1 and then to M_3, a front-silvered mirror. The transmitted portion of the origi-nal beam passes through the compensator plate C, a plate of the same material as M_1 but without the silvering. With C, both optical paths are equivalent if the distances to M_2 and M_3 are the same when measured from the point at which the beam is split. Mirror M_2 is also a front-surfaced mirror.

If the path lengths are the same, the observer viewing the mirrors M_2 and M_3 through his telescope will see a bright fringe. If the path lengths are not the same, the phase difference will determine whether the fringe observed at the center of the tele-

---

[1] A half-silvered mirror is a mirror on which the silver coating is applied so that only 50% of the light striking the silver surface is reflected. The remainder is transmitted. The half-silvering assures us that the relative intensity of the light in each path is about equal.

scope T will be bright or dark. The field as viewed through the telescope is shown in Figure 2.7. Vertical straight line fringes are seen as the images of the slit S. In the absence of S, the fringes as viewed in the telescope appear round as the images of the lens.

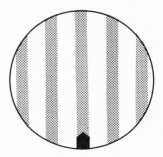

Figure 2.7.   The image presented in the telescope of a Michelson interferometer.

Mirror M_2 is a fixed mirror which can only be adjusted for tilt so that it can always be fixed normal to the beam, while M_3 has a mounting on a screw in addition to the tilt adjustment. This screw is a carefully machined micrometer screw so that M_3 can be moved parallel to the beam and the distance of movement accurately measured.

As M_3 is moved either closer to or away from M_1, the path difference, and therefore the phase difference, between the two optical paths varies, and the observer sees the fringes move across the field of the telescope. The telescope is provided with a small pointer, as in the figure, or with cross-hairs to provide a point of reference for the observer. As M_3 is moved, the observer counts the number of fringes which cross the field. Movement of M_3 a distance of $\lambda/2$ changes the path length in that arm of the interferometer by λ since the beam travels the distance $M_1 M_3$ twice (once to and once from M_3). A change of $\lambda/4$ in the position of M_3 will then move the dark fringe shown in Figure 2.7 either to the right or left one place, and replace it with a bright fringe. The distance M_3 moves divided by the number of fringes crossing the pointer of the telescope will equal $\lambda/4$.

Example 2.4. A Michelson Interferometer is used to measure the wavelength of yellow sodium light. The moveable mirror is found to have been shifted a distance of 1.47×10^{-2} mm when 100 fringes have been counted.

Using our equation with N equal to the number of fringes and d equal to the distance of travel of the moveable mirror we get:

$$N \times \frac{\lambda}{4} = d$$

$$\frac{100}{4} \lambda = 1.47 \times 10^{-2} \, \text{mm}$$

and $\lambda = 589 \, \text{m}\mu$ which is consistent with what we expect for yellow light.

Thin Film Interference. Another class of interference experiments was actually known before Young's experiment was performed. This is the interference resulting from reflections in thin films. The color seen in soap bubbles is a result of this kind of interference.

Robert Hooke (1635–1703) first observed the fringes that result when two spherical surfaces of unequal curvature are placed in contact. Sir Isaac Newton (1642–1727) studied the effect in more detail and the fringes bear the name *Newton's rings*. They may be observed by placing a convex lens in contact with a plane glass surface. Figure 2.8 illustrates the arrangement of the experi-

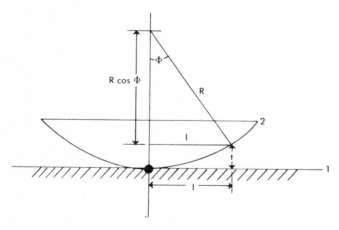

Figure 2.8.   Experimental set-up for the Newton's rings experiment.

ment. Figure 2.9 shows the view seen by an observer looking down on this arrangement.

The interference that is seen requires the superposition of two waves. These waves arise because the light entering the lens from

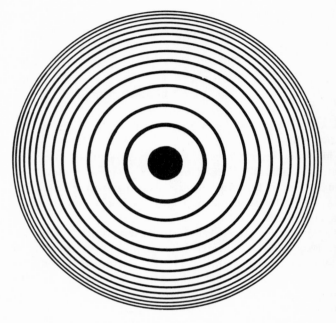

Figure 2.9. A normal view of Newton's rings.

above is partially reflected by the underside of the lens (surface
in the figure) and partially reflected at the plane plate (surface 1
These two reflected waves are the waves which interfere and th
additional path length, 2t, of the wave reflected at surface 1 giv
the potential for phase changes.

A most interesting aspect of the experiment is seen in Figur
2.9. The center of the interference pattern is dark, correspondin
to a phase change of $\lambda/2$ or 180° between the wave reflected
surfaces 1 and 2, in spite of the fact that these surfaces are i
contact at their centers. The zero path difference at the conta
point coupled with the appearence of interference suggests tha
in the process of reflection some change in phase must occu
From electromagnetic theory it is known that when a light wav
undergoes reflection at the surface of a more optically dense me
dium (a medium with a higher index of refraction), a phase shi
of 180° is introduced into the reflected ray. If the light wave i
reflected by a less dense medium, no phase change occurs. Th
immediately explains the result of Newton's experiment. At th
center, the wave which is reflected at surface 2 is reflected at

glass-air interface and therefore undergoes no phase change, while the wave reflected at surface 1 comes from the very thin layer of air between the two surfaces and is reflected at the glass with an air-glass or a rare-dense reflection so that a 180° phase change occurs. The net result is that the two waves traveling back toward the source from the two reflections are 180° out of phase and cancelled. The center of the pattern is black.

The remaining black circles occur at various radii from the center of the lens. In order to find the radius r at which a given dark interference occurs, we must resort to an approximation based on Figure 2.8. The thickness of the film at distance r from the center is given by:

$$t = R - R \cos \Phi = R(1 - \cos \Phi)$$

where the quantities used are defined in the figure. We make two other important approximations: (1) since Φ is small we can use the first two terms of its cosine series expansion in its place:

$$\cos \Phi = 1 - \frac{\Phi^2}{2}$$

and (2) since Φ is small we can replace it by the angle itself, given in radians:

$$\Phi = \frac{r}{R}$$

Combining these equations we get:

$$t = \frac{r^2}{2R}$$

We will see destructive interference whenever 2t is an integral wavelength multiple, since the additional $\lambda/2$ required for destruction is provided by the phase change in the reflection. This means that the condition for a ring of interference to be dark is:

$$2t = \frac{r^2}{R} = m\lambda$$

where m is zero or some integer. The dark radii then appear at:

$$r = \sqrt{mR\lambda} \qquad (2.2)$$

Example 2.5. Yellow light falls normally upon a plano-convex lens, with a radius of curvature of 60 cm, resting on a flat glass plate. The light is 590 mμ in wavelength. At what radius does the fortieth dark ring occur. (Do not count the central dark spot.)

This problem may be solved directly using equation 2.2.

$$r = \sqrt{40 \times 60 \times 590 \times 10^{-7}} = 0.376 \text{ cm}$$

Example 2.6. Thin wedges of air also produce the same effect as the lens and flat glass plate of Newton. Two glass microscope slides, 10 cm long, which touch at one end and are held apart at the other end by a thin sheet of paper will show interference patterns as shown in Figure 2.10. If the paper separating the two slides is 0.01 cm thick, how many dark fringes will be seen when the slides are viewed in 700 mμ light?

Figure 2.10. A thin wedge of air between two microscope slides.

There will be a dark fringe at the point of contact of the two, and another each time the distance between the two slides increases by one-half wavelength. At the point of support, the distance between the two slides is 0.01 cm or 10^5 mμ. One-half a wavelength is 350 mμ. Dividing, we can see that there are 285 half-wavelengths. (Anything less than a full half-wavelength is neglected since it cannot lead to another fringe.) We then will see 285 fringes, one for each half-wavelength increase in separation of the plates, plus the dark fringe at the point of contact. Altogether, there will be 286 dark fringes.

Closely related to the interference effects of thin air films just discussed is the interference resulting from thin films of various materials such as soap bubbles and oil films on wet pavements. The colored appearance of these films is due to interference effects in the white light illuminating them. We will see, however, that some modification in our previous formulation is necessary.

Figure 2.11 illustrates a thin film of any material illuminated from above. We will consider the illumination to be derived from a white source and to be normally incident on the film (although for purposes of illustration we have drawn the incoming rays obliquely). The film is of thickness t and has air both above and below, as in the case of a soap bubble. The light striking the film along path AB is partially reflected and partially transmitted. The transmitted portion suffers partial reflection at point

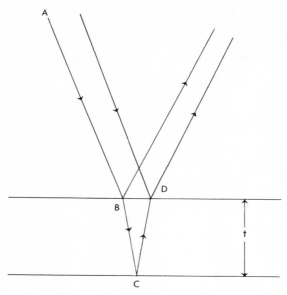

Figure 2.11. Reflection paths in a thin film.

C, and returns to point D to recombine with the wave reflected at the upper surface. We can determine the interference effects by determining the relative phases of the two waves as they recombine. A phase shift will be introduced by the additional path BCD traveled by the partially transmitted portion of the incident wave. The phase shift will be determined by the number of waves in the distance 2t. Within the soap film the wavelength will not be λ_0, the free space of air wavelength, but will be modified because the index of refraction of the material modifies the wavelength. In a material medium, the velocity of light is reduced over the free space velocity, and the ratio of free space velocity to the velocity in the medium is the index of refraction, n. Since the frequency of light is unaltered by changes in medium, the wavelength must also decrease to λ_0/n in the medium.[1] We

---

[1] The equation of wave propagation reads:

$$c = \lambda_0 \nu$$

where c is the velocity of light in free space and ν is the frequency. Since $n = c/v$ in the medium, c is replaced by c/n. To maintain the equality λ_0 must also be divided by n. The wavelength in the medium is therefore λ_0/n.

must also remember that there is a 180° (π radians), or $\lambda_0/2$, phase change in the ray reflected at point B (and D, etc.) because this is a rare-to-dense reflection. The total phase shift, Φ, which occurs is then:

$$\Phi = 2\pi \frac{2t}{\lambda_0/n} - \pi \qquad (2.3)$$

If Φ is 2π, or some multiple of 2π, there will be constructive interference in the reflected ray. If the shift is π, or some odd multiple of π, the reflection will interfere destructively. A number of examples will illustrate the use of equation 2.3. In general, the best procedure in solving problems of this class is to derive the required relationship from the physics of the problem rather than to rely on some memorized equation.

Example 2.7. A soap bubble appears red in reflected light. (Take red light to have a wavelength of 610 mμ, as in Appendix 1.) The index of refraction of the bubble is 1.35. How thick is the bubble?

This problem can be solved directly with equation 2.3. The phase shift must be some multiple of 2π. If we solve 2.3 for t we get:

$$t = \frac{\lambda_0}{2n}(m + \tfrac{1}{2}) \qquad (2.4)$$

Solving this equation gives us:

$$t = 226(m + \tfrac{1}{2})$$

where m is some integer or zero. For the lower order interference, $m = 0$ and t = 113 mμ. There is no way we can determine the order of the interference directly so that m may equal one and t equal 339 mμ (or $m = 2$ and t = 565 mμ, etc.). Suppose that $m = 0$ so that t = 113 mμ. Is there another wavelength in the visible region that can interfere constructively?

The question can be restated in terms of λ_0 and m by asking if a λ_0 and an m can be found so that λ_0 is in the visible region of 400–700 mμ. If we solve equation 2.4 for λ_0 in terms of m, we get:

$$\lambda_0 = \frac{2.7\,t}{m + \tfrac{1}{2}}\,m\mu$$

where the known quantities have been substituted. With $m = 0$, we get $\lambda_0 = 610$ mμ; with $m = 1$, we get $\lambda_0 = 20.3$ mμ; further

ncreases in m lead to a decreasing λ_0. Therefore, there will be
one other (and only one other) wavelength in the visible region
which will give constructive interference. That wavelength will be
410 mμ in the violet region, and the bubble will appear reddish
n reflection.

Example 2.8. A thin soap bubble of thickness 165 mμ is seen
n transmitted light. The index of refraction of the bubble is 1.35.
What color will it appear in transmitted light?

This problem is different from the others we have treated in
the section on material films. Figure 2.12 shows the situation for
transmitted light in this case. The interference is determined by

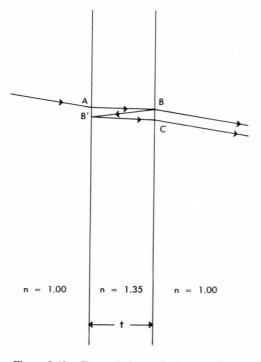

Figure 2.12. Transmission paths in a thin film.

the phase difference between the ray exiting the film at B and the
ray exiting the film at C. The additional path length traveled by
the ray exiting at C is 2t. All reflections are of the dense-to-rare
variety and involve no phase changes. The phase change, Φ, is

then given by:

$$\Phi = \frac{2t}{\lambda_0/n} \, 2\pi \qquad (2.5)$$

In order for there to be constructive interference, Φ must be a multiple of 2π. We then must find for each value of m the possible values of λ_0 which give constructive interference in the visible region.

Equation 2.5 when solved for λ_0, with Φ replaced by $m2\pi$ gives:

$$\lambda_0 = \frac{2nt}{m}$$

With m equal to one and the appropriate numbers substituted from the problem statement, we find λ_0 equals 446 mμ, a value in the violet region of the spectrum. Any larger values of m will only decrease the value of λ_0, and thus the film appears violet in transmission.

Lenses of cameras and binoculars and other optical instruments are coated with a thin film of material to minimize the light loss through reflection. This gives rise to the purplish hue of lenses in these instruments. The coating is generally made of such a thickness as to eliminate the reflection of yellow-green light at 560 mμ. It is impossible to provide a coating that will eliminate all the reflections in the visible because of the variability of the wavelength of visible light, but 560 mμ is chosen because it is the wavelength of the maximum sensitivity of the eye. The coating is chosen with an index between that of air and the optical glass so that the amount of light reflected at the glass-material interface is about equal to that reflected at the air-material interface. Figure 2.13 shows the situation with $n_{glass} = 1.55$ and $n_{coating} = 1.40$.

From the figure, we see that both reflections are rare-to-dense. Therefore, the reflection at the coating material and at the glass each involve a phase change of π radians, and these compensate so that we need not consider them. For *destructive* interference at point C, the path ABC must be one-half wavelength longer. There are, of course, higher orders of destructive interference which can take place at the interface, but the minimum thickness of the coating is usually taken. ABC must be one-half wavelength longer as modified by the coating material and:

$$2t = \frac{1}{2}\left(\frac{\lambda_0}{n_{coating}}\right)$$

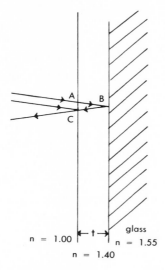

Figure 2.13. A thin film lens coating.

from which we can calculate $t = \lambda_0/4n_{coating}$. With the values
which are typically used, as given above, $t = 100$ mμ.

To find the color which is constructively reflected, we note
that the thickness of the film must be $\lambda_0/2n_{coating}$ from which λ_0
is seen to be 280 mμ in the violet region beyond the visible but
close enough to give the lens a violet cast.

Other Types of Interferometers. Two other types of interferometers are of importance in modern technology. These are the
Mach-Zender and the Fabry-Perot interferometers.

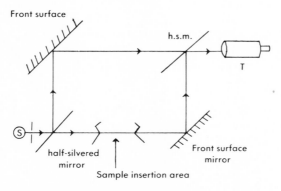

Figure 2.14. The Mach-Zender interferometer.

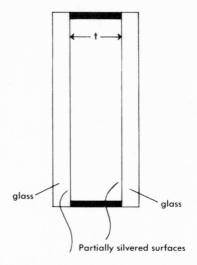

Figure 2.15. The Fabry-Perot interferometer.

The design of the Mach-Zender instrument is illustrated in Figure 2.14. The half-silvered mirror acts as a beam divider and the optical paths are adjusted until they are equal. Into one arm of the interferometer, a material such as a gas under pressure is introduced. This modifies the optical length of the arm, and the observer sees fringes sweeping past his telescope. By counting fringes, the change in the optical path can be determined, and thus the index of refraction of the material introduced can be determined quite accurately.

The Fabry-Perot Interferometer consists of two partially silvered plates set a known distance apart, as shown in Figure 2.15. The distance between the plates is fixed so that only a single narrow range of wavelengths is transmitted. Such an interferometer is often called an interference filter, and these filters are commercially available for various purposes.

Chapter 3
Diffraction

The basis for the ray model of light and for the ideas associated with rectilinear propagation of light came from the ob-
servation of sharp shadows which occur when a screen is illuminated with light from a point source. This is illustrated in Figure
3.1 where a point source is chosen so as to eliminate a penumbra.
Actually the situation is much more complex. Figure 3.2 shows
the detailed distribution of the intensity of the light in the region
near the edge of the shadow.

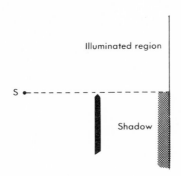

Illuminated region

S

Shadow

Figure 3.1. The geometric shadow cast by an opaque object under the assumption of rectilinear propagation of light.

Effects such as are shown in Figure 3.2 are known as diffraction effects and they are seen whenever an object in the path of a
wave front acts to limit the size of the front. One does not observe
diffraction effects very often because most natural sources are
extended sources and the diffraction pattern produced by various
parts of the source overlap and statistically cancel.

Diffraction of light is often thought of as the bending of light
around obstacles. This is not strictly correct since diffraction
occurs naturally as a result of the continued propagation of that
portion of the wavefront which has not been cut off by the inter-
posed object. The main principle which will aid us in understanding diffraction is *Huygen's principle* which states that every point

33

Figure 3.2. The diffraction pattern resulting from the shadowing of a straight edge as shown in the previous figure. The drawing is on an expanded scale to show the detail. The intensity of the light in the region near the geometric shadow is twice that in the open region to the left.

of a wave front can be considered as a secondary source from which waves propagate in all directions.[1] If we consider the wavefront above the opaque object in Figure 3.1 we can see that secondary waves can propagate downward into the shadow region. This chapter will be devoted to the discussion of diffraction.

The Single Slit. The single slit is one of the most graphic examples of diffraction effects. If we allow light from a point source to pass through a slit of variable width as shown in Figure 3.3, a wide opening, *a*, gives rise to simple shadowing. As the width of the slit is reduced, the shadow region takes on a fringed appearance, *b*. The beam, after passing through the slit, spreads out to give a diffraction pattern consisting of a bright central band; then bright and dark fringes alternate with a rapidly decreasing intensity. One can easily see this effect by observing a light source between his fingers with the opening made progressively smaller and smaller.

It is easy to see how this occurs on the basis of Huygen's principle. Figure 3.4 shows the slit and the secondary waves which can be thought of as propagating from each point in the wave front at the slit. The intensity at each point on the screen is then determined by superposing each of the waves which arrive from the secondary sources. There will of course be phase differences due to the different distances from each of the secondary sources to any fixed point on the screen. In addition, the mathematical development of Huygen's principle leads to an *obliquity factor*, or *inclination factor*, which causes the secondary waves which are propagated at an angle of Φ to the original wave to

[1] For a discussion of Huygen's principle see *Optics I*.

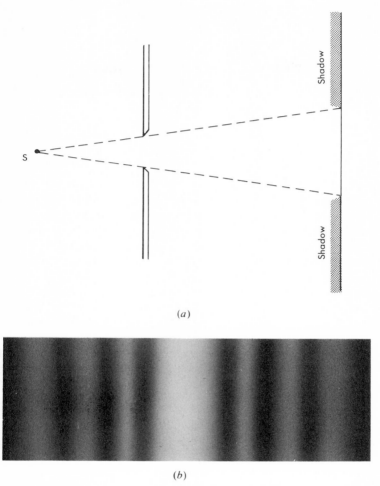

(a)

(b)

Figure 3.3. (a) Shadowing by a wide slit. (b) Diffraction by a narrow single slit.

have an intensity proportional to $\frac{1}{2}(1 + \cos \Phi)$. This factor leads to a maximum intensity of one for waves propagated in the direction of the original wave, and zero for waves propagated at right angles to the original wave. It is the obliquity factor which causes the rapid decrease in the intensity of the diffraction pattern as you move away from the central maximum. The intensity of the bright band adjacent to the central band is 0.05 times as intense as the bright band.

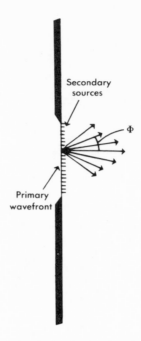

Figure 3.4. The details of the single slit showing the primary wavefront, the constructed secondary sources, and the rays propagating from one of the secondary sources. The angle Φ is the angle which appears in the obliquity factor.

There are two particular cases of diffraction which we will treat. These are limiting cases, and they merge in the region between the limits. The first limit is that case where the rays reaching the screen are paraxial. This case occurs when the screen is relatively far from the slit, or the slit is very narrow. It is known as *Fraunhofer diffraction*. The second limit is where the slit is relatively wide, or the observing screen close to the slit, so that the paraxial assumption does not hold. This case is called *Fresnel diffraction*. Fraunhofer diffraction also occurs when a lens is placed between the slit and the screen in order to effect the convergence of the light from the slit and make the light passing through the slit parallel.

Fraunhofer Diffraction.   The case of Fraunhofer diffraction by a single slit is one of the simplest cases of diffraction which we can treat. The apparatus used is shown in Figure 3.5. Light from the source is rendered parallel by the lens L_1. After passing

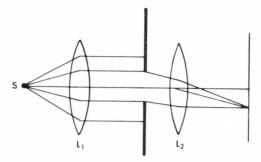

Figure 3.5. The experimental set-up used in observing the Fraunhofer diffraction pattern of a single slit.

through the slit the light is focused on the screen by L_2. The diffraction pattern observed is essentially the one shown in Figure 3.3b, with a broad central maximum and narrower bright and dark fringes adjacent to it. The width of the central maximum is about twice that of the other fringes. As the slit is made smaller, the width of the central maximum increases showing that the pattern is strongly dependent upon the slit width, w.

One of the properties of lenses suggests that only plane waves reaching L_2 will be focused on the screen. We can therefore make a construction (like Figure 3.6) of the light leaving the slit, and can base our analysis merely on the difference in path length traversed by various secondary rays originating at the slit.

Consider the wave leaving the slit at an angle of Φ with the direction of the plane wave from the primary source. The region on the screen where the center line CC strikes will be the central

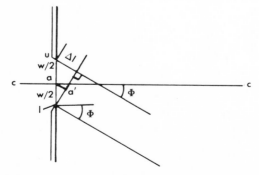

Figure 3.6. Construction for the Fraunhofer diffraction by a single slit.

maximum. Suppose that Δl is one wavelength, λ, in length. From the geometry of Figure 3.6, we can see that the short centerline aa' must be $\frac{1}{2}\lambda$ in length. If we consider secondary waves with their origin at the bottom of slit L and with their origin at point a, both moving toward the screen at an angle of Φ, we can see that they will be 180° out of phase (corresponding to $\frac{1}{2}\lambda$ path difference) when they reach the screen and they will then cancel. As we consider each point from L to a along the lower half of the slit, there is a corresponding point from a to U which emits a secondary wave with a 180° phase shift. The result will be that at the screen all arriving waves at the angle Φ will cancel and we will have a minimum.

Figure 3.7 shows the geometry including the screen. From the

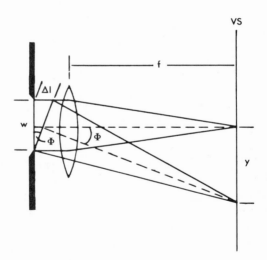

Figure 3.7. Viewing screen and lens for the Fraunhofer diffraction experiment.

geometry of the figure we can see that:

$$\sin \Phi = \frac{\Delta l}{w} \approx \frac{y}{f} \qquad (3.1)$$

We have seen that a Δl of λ results in a minimum. Now suppose that Δl is $\frac{3}{2}\lambda$. In this case, we can divide the slit up into three zones as in Figure 3.8. Point by point, there is phase cancellation of the light reaching the screen from the zone ab and the zone bc

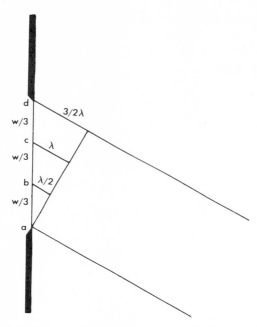

Figure 3.8.   Three zones used in the determination of the first maximum in Fraunhofer diffraction.

while the zone *cd* will not have a zone which will compensate with it. The result will be that when Δl is $\frac{3}{2}\lambda$ we will have a maximum in the diffraction pattern.

This maximum will of course be considerably less intense than the central maximum since light from only one-third of the slit illuminates it and, in addition, the obliquity factor serves to diminish the intensity of the light going to this region. It is clear from the two cases just treated that maxima will occur whenever

$$\Delta l = \left(n + \tfrac{1}{2}\right)\lambda \tag{3.2}$$

where n is some nonzero integer. Minima, on the other hand, will occur whenever:

$$\Delta l = n\lambda \tag{3.3}$$

where n again is some nonzero integer.

The approximate relative amplitude of the light at some maximum is the fraction of the slit which contributes to the illumination (the obliquity factor is omitted). In the case treated above, this will be one-third.

The approximate relative intensity will be the square of the relative amplitude, since intensity is always derived from the square of the amplitude. In the above case, this will then be one-ninth, and the central maximum will be nine times brighter.

Example 3.1. A Fraunhofer diffraction experiment is performed with 600 mμ light and a lens with a focal length of 50 cm. Find the position of the second maximum and estimate the relative intensity if the slit width is 0.1 mm.

In order to solve this problem we first note that the first maximum occurs with n = 1, the second with n = 2, etc. Using equation 3.2 we can find Δl.

$$\Delta l = \left(2 + \tfrac{1}{2}\right) \times 600 \text{ m}\mu = 1500 \text{ m}\mu$$

This result can be used now in equation 3.1 to find the first result required, namely y.

$$y = f \frac{\Delta l}{w} = 50 \text{ cm} \times \frac{1.5 \, \mu}{100 \, \mu} = 0.75 \text{ cm}$$

The second maximum is displaced 0.75 cm from the central (or zeroth) maximum.

In order to evaluate the relative intensity we can consult Figure 3.9. This figure is drawn for the situation of this problem. We can see that in the case of a second maximum there are five zones and therefore one zone which does not cancel. One estimate of the relative intensity is therefore, $\left(\tfrac{1}{5}\right)^2$ or $\tfrac{1}{25}$ of the intensity of the central maximum.

A better estimate would include the obliquity factor. We can evaluate the angle needed for the obliquity factor from equation 3.1.

$$\sin \Phi = \frac{\Delta l}{w} = \frac{1.5 \, \mu}{100 \, \mu} = 0.015$$

and the angle Φ is 0.86°. The cosine of this angle is so close to unity that very little change is made in the estimated intensity. If we want to include the obliquity factor, we must remember that it also acts on the amplitude. When including it in the intensity, we must remember to multiply by the square of this factor.

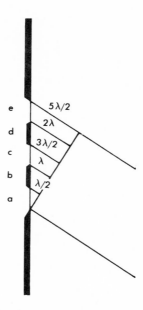

Figure 3.9. The zones for the second maximum of a Fraunhofer diffraction problem. Zones a and b pair with c and d to cancel, leaving one-fifth of the slit as the illuminating area.

Fraunhofer Diffraction at Circular Apertures. The calculation for the diffraction pattern due to a circular aperture is essentially the same as for a slit. In this case, however, the details of the calculation are much more complex because of the circular geometry. It is an important case because most lenses, including the eye, are circular in cross-section, and thus form images which have the diffraction characteristics of a circular aperture.

The diffraction pattern appears as a bright central disc with alternating bright and dark rings bordering it. The intensity of the central disc is, as we have seen before, much greater than for the secondary maxima. Equation 3.1 relates the angle subtended at the lens for slit patterns to the slit opening and the wavelength.

The first minimum in the diffraction pattern formed by a circular aperture is given by:

$$\sin \Phi = 1.22 \frac{\lambda}{d} \tag{3.4}$$

where d is the diameter of the circular opening.

Example 3.2. The lens on a monocular is 4.7 cm in diameter and has a focal length of 20 cm. How large is the image of a point object at infinity? Assume that the light has wavelength 560 mμ.

Using equation 3.4 we have:

$$\sin \Phi = 1.22 \frac{5.6 \times 10^{-5}}{4.7} = \frac{y}{f} = \frac{y}{20 \text{ cm}}$$

and the radius of the image is found to be 2.90×10^{-3} mm. This is small, as one expects from the large diameter of the lens and the short focal length. The image formed by a large astronomical telescope may be as large as a millimeter, however.

If we observe two point sources which are very close together and at a large distance, it is conceivable that, because of the size of the diffraction discs of their images, they may run together and be indistinguishable. A condition known as *Rayleigh's criterion* has been fixed for the determination of resolvability of point sources. This condition states that two objects can be resolved if the maximum of one pattern is no closer to the second pattern than the first minimum. Figure 3.10 is a plot of two diffraction intensities at the viewing screen with the separation chosen so as to just fulfil Rayleigh's criterion.

From equation 3.4 we can see that the necessary separation of the images on the viewing screen is:

$$y = \left(1.22 \frac{\lambda}{d}\right)(f)$$

where f is again the focal length of the lens. The angular separation of the images is $1.22 \frac{\lambda}{d}$. (Here we have assumed that $\sin \Phi = \Phi$ which is true for the very small angles involved.)

Whenever two sources are to be observed, they will be resolved if their angular separation subtends an angle greater than α at the lens or slit where the diffraction pattern is produced where:

$$\alpha = 1.22 \frac{\lambda}{d} \tag{3.5}$$

according to Rayleigh's criterion. This relationship is particularly valuable to astronomers who need to resolve distant point sources. Since the wavelength of the light is essentially fixed, the resolution can best be improved by increasing the diameter of the aperture. The largest lens aperture is 40 inches at the Yerkes

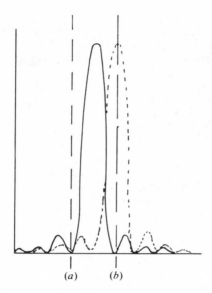

(a) (b)

Figure 3.10. The intensity of the diffraction pattern at a viewing screen for an image of a distant point source formed by a circular aperture and lens. A second maximum is shown centered on point b which is the limit point for resolvability according to Rayleigh's criterion. Any other source which was close enough to the source giving the full line pattern above, so that its maximum fell between a and b on the viewing screen, would not be resolved.

Observatory, and any of the larger telescopes are built with imaging mirrors up to 200 inches. The mirror in a reflecting telescope functions as the aperture in the same way as a lens by limiting the portion of the wavefront which is observed.

Example 3.3. What is the limit of angular resolution for the Yerkes 40 inch telescope?

We will use 560 mμ (= 5.6×10^{-5} cm) as the wavelength. (40 in. = 100 cm.) Using equation 3.5 we see that:

$$\alpha = 1.22 \frac{\lambda}{d} = 1.22 \frac{5.6 \times 10^{-5}}{100} = 6.85 \times 10^{-7} \text{ radians}$$

which corresponds to 0.141 seconds of arc.

We can compare the above value to the eye, where the effective diameter of the pupil is perhaps 2 mm.

$$\alpha_{\text{eye}} = 1.22 \frac{5.6 \times 10^{-5}}{0.2} = 3.4 \times 10^{-4} \text{ radians}$$

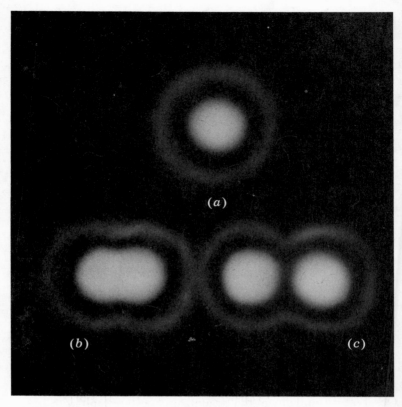

Figure 3.11. (a) The Fraunhofer diffraction pattern produced by a circular opening. (b) Two point sources at the limit of resolution by the Rayleigh criterion. (c) Two well resolved point sources.

which corresponds to about one minute of arc so that the telescope has 1000 times better angular resolution. The eye's resolution is better at night when the pupil opens to admit more light, but the increase is only by a factor of two or three.

Fresnel Diffraction. We are dealing here with the case of a relatively large aperture or a screen close to the aperture. The approach we will use is one which was developed by Fresnel to handle this case where the rays are not paraxial. We will develop this treatment in only a semiquantitative way since the full mathematical treatment involves the evaluation of some difficult integrals. What we will do is divide the wavefront up into zones of finite (rather than infinitesimal, as in the exact theory) width and then superpose the waves from each of these *Fresnel zones*.

Figure 3.12 shows the experimental arrangement. Light from a distant source falls on a circular opening of radius r. The wavelength of the light is taken to be λ. The waves striking the screen

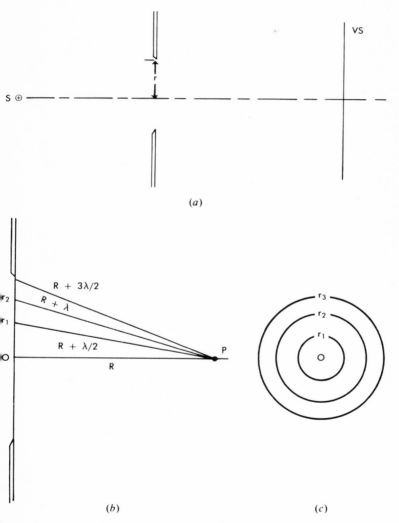

(a)

(b)   (c)

Figure 3.12. The experimental arrangement (a), the construction of the Fresnel zones (b), and the Fresnel zones (c) for a circular aperture.

are essentially plane waves. As in Figure 3.12b, we take P to be the intersection of the central axis with the screen, and from P we lay off circles in the plane of the slit with radii $R + \frac{1}{2}\lambda$, $R + \lambda$, $R + \frac{3}{2}\lambda$, etc. These circles are generated by the circular symmetry about OP in Figure 3.12b. The wave surface at the aperture is then said to be divided into *half-period zones* or *Fresnel zones*, each of which has the property that the distance to P increases by one-half wavelength as one passes from the inner to the outer edge of the zone. The zones themselves can be seen in Figure 3.12c.

The computation of the radii of the zones is straightforward through the Pythagorean theorem. For the first zone:

$$R^2 + r_1^2 = \left(R + \frac{\lambda}{2}\right)^2 = R^2 + R\lambda + \left(\frac{\lambda}{2}\right)^2$$

which reduces to:

$$r_1^2 = R\lambda$$

where we have dropped the term in λ^2 because it is at least 10^{-5} times smaller than any of the other terms in the equation. The radii of the zones is then given by

$$r_1 = \sqrt{R\lambda}$$
$$r_2 = \sqrt{2R\lambda} = \sqrt{2}r_1 \qquad (3.6)$$
$$r_n = \sqrt{nR\lambda} = \sqrt{n}r_1$$

where we have used the same procedure for successive zones and then made the extension to the general case by mathematical induction.

If we take $\lambda = 560$ mμ and $R = 1$ m then:

$$r_1 = (560 \times 10^{-9} \times 1)^{\frac{1}{2}} = 7.5 \times 10^{-4}\,\text{m} \approx 0.75\,\text{mm}$$

The radii of the zones always is in the ratio: $\sqrt{1}$, $\sqrt{2}$, $\sqrt{3}$, etc., but r_1 depends on both the wavelength of light and the distance to the viewing screen.

In order to determine the amplitude to the disturbance at P, we have to use the principle of superposition on the disturbances from each of the zones. As we go from zero to r_1, the phase of the wave at P goes from zero to π. From r_1 to r_2, the phase goes from π to 2π. The waves from the second zone will cancel the waves from the first zone. Following this argument out with successive zones we find that the even-numbered zones are out of phase with the first zone, while the odd numbered zones are in phase with it. Successive zones will tend to cancel each other.

We must also determine the illumination amplitude from each zone in order to follow through with the superposition. The amplitude reaching P from each zone will be proportional to the area of the zone. The area of the first zone is:

$$A_1 = \pi r_1^2$$

The area of the second zone is:

$$A_2 = \pi r_2^2 - \pi r_1^2 = 2\pi r_1^2 - \pi r_1^2 = \pi r_1^2$$

That of the third zone is:

$$A_3 = \pi r_3^2 - \pi r_2^2 = 3\pi r_1^2 - 2\pi r_1^2 = \pi r_1^2$$

etc. The areas of all the zones are equal so that the amplitude contribution of each zone will be the same and the disturbance at P will be only dependent on the number of zones and their phase relationship.

The total disturbance at P can then be given by:

$$D = D_1 - D_2 + D_3 - D_4 + D_5 - \cdots \tag{3.7}$$

Where D is the total disturbance and D_n is the disturbance due to the nth zone. The signs alternate because the zones are alternately in and out of phase.

Actually the amplitudes decrease slowly as we go from the first to the nth zone. The amplitude of the nth zone is generally taken as:

$$D_n = \frac{D_{n+1} + D_{n-1}}{2}$$

If we substitute this back into equation 3.7 and perform the infinite sum we find that:

$$D = \frac{D_1}{2}$$

and the disturbance at P with the entire wavefront unobstructed is only one-half as great as with the first zone alone.

We can go back to Figure 3.12 and determine the disturbance at P. There are just three zones in the aperture. The first and second cancel and the third creates the total disturbance which is somewhat less than D_1. Enlarging the slit decreases the intensity of the disturbance at P because the fourth zone which will now be exposed will cancel the third. When the fifth zone comes into the aperture as the aperture is opened, the intensity will again increase.

Decreasing the radius of the aperture will decrease the illumination until the slit radius gets to r_2, at which point D at P will be zero. Further decrease in the radius of the slit will cause the illumination to increase until the radius is r_1, at which point the illumination at P is at its maximum value.

If P is off the axis, as in Figure 3.13a we get the Fresnel zone picture shown in Figure 3.13b. The determination of the disturbance at P is difficult, but a very good approximation may be had by adding together the areas of the even zones and subtracting the odd zone areas. A result of zero means no illumination or disturbance at P, while a number close to πr_1^2 means a large illumination.

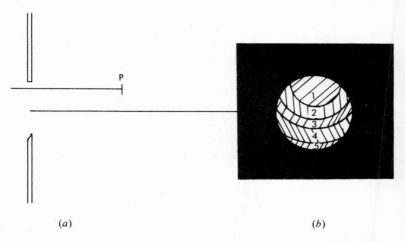

(a)   (b)

Figure 3.13. An off axis point (a) and the Fresnel zones for the point (b).

Two interesting devices come from our analysis of Fresnel diffraction. A *zone plate* is a flat piece of transparent material in which the even-numbered zones have been blacked out. Such a plate is shown in Figure 3.14. The result of eliminating the even zones is to place all the disturbance at the focal point in phase. There is a focal length for a zone plate because the zone radii must be constructed for some set distance from the plate to the

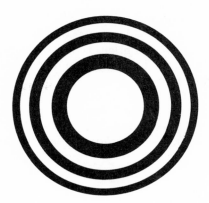

Figure 3.14. A circular zone plate drawn somewhat oversize. The focal length for this plate would be about 180 m.

screen and for some wavelength, as in equation 3.6. A zone plate will produce tremendous heat when used to focus the sun's rays and can be used to kindle fires as easily as a pocket lens (and generally more efficiently).

The other interesting application of our analysis is the Fresnel lens used in theatrical lights, where very intense illumination is required. These lenses are most generally used for fixed spot-lights and are shaped as shown in Figure 3.15. The object of the stepped thickness is to shift the phase of the successive zones into phase with the first zone using the changed optical thickness of the glass to effect the phase shift.

For a circular obstacle rather than aperture, several of the central zones are covered and the intensity is determined by summing, as in the case of the full wavefront, the contribution of the remaining zones. The illumination at the center under a circular obstacle turns out to be nearly the same as the illumination without the obstacle. The pattern then turns out to be a bright

Figure 3.15. The Fresnel lens.

central spot surrounded by dark and relatively dark fringes out to the edge of the geometric shadow (after which the pattern tends to become evenly intense). The bright central spot can be seen if a circular object is placed before the sun and photographed.

Fresnel diffraction by a straight edge results in the pattern shown in Figure 3.2. The zones in this case are not circular, but are narrow strips parallel to the edge of the obstacle, as shown in Figure 3.16. The distance from O to the zone edges is found again by equation 3.6. We will not consider this case in any detail.

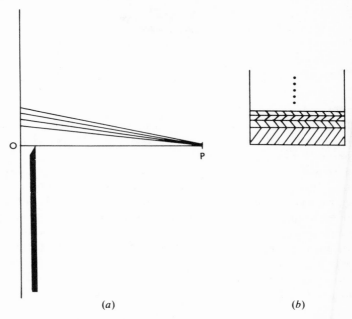

(a)   *(b)*

Figure 3.16. (*a*) The Fresnel construction for diffraction by a straightedge and (*b*) the zones.

Two examples will help to demonstrate the calculations involved in using the Fresnel diffraction analysis.

Example 3.4. Light of wavelength 600 mμ falls on a circular orifice 0.3 mm in diameter. Where on the axis of the orifice will the intensity of the light be a maximum?

What we require here is the distance at which the first zone just fills the aperture. From equation 3.6:

$$r_1 = \sqrt{R\lambda} \qquad \text{or} \qquad r_1^2 = R\lambda$$

We see that:

$$R = \frac{r_1^2}{\lambda} = \frac{(1.5 \times 10^{-4})^2}{600 \times 10^{-9}} \approx 0.0375 \text{ m} \approx 3.75 \text{ cm}$$

Then at 3.75 mm the intensity is a maximum. The intensity on the axis will fall to a minimum when the first two zones fill the aperture. This distance can be found using equation 3.6, and is equal to 1.87 cm.

The analysis is not limited to light waves as can be seen from the following example.

Example 3.5. A loudspeaker designed to broadcast sound of wavelength 1 ft has a cabinet opening of radius 1.5 ft. Where will the sound have greatest intensity?

The opening must just enclose the first Fresnel zone. Once again using equation 3.6, we see that:

$$r_1^2 = R\lambda$$

$$R = \frac{r_1^2}{\lambda} = \frac{(1.5)^2}{1} = 2.25 \text{ ft}$$

The sound is most intense at 2.25 ft from the speaker directly in front of it.

Example 3.6. Monochromatic light of wavelength 563.3 mμ passes through a circular opening 1.3 mm in radius. If the Fresnel diffraction pattern is observed on a screen 1 m from the opening, is the central spot light or dark?

We need to know how many zones are included in the opening. From equation 3.6:

$$r^2 = n\lambda R$$

$$n = \frac{r^2}{\lambda R} = \frac{(1.3 \times 10^{-3})^2}{563.3 \times 10^{-9} \times 1} = 3$$

Three zones are uncovered and the central spot is bright.

We have treated Fresnel and Fraunhofer diffraction as distinct. Actually, there is a point at which they are no longer distinct, but merge into one. Fresnel diffraction analysis is used when the screen is close to the diffracting object and/or the object is large, while Fraunhofer techniques are used when the rays leave the orifice paraxially or when the orifice is small.

The Diffraction Grating. The diffraction grating combines a problem in diffraction with a problem in interference. The grating consists of a number of closely spaced slits ruled into some opaque material. The light which passes through the slits falls on a condensing lens and is focused on a viewing screen. The entire viewing arrangement is shown in Figure 3.18.

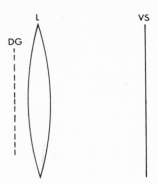

Figure 3.17. The diffraction grating with condensing lens and screen.

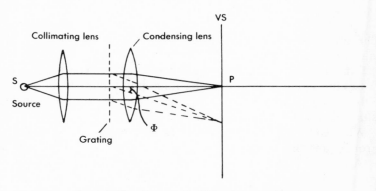

Figure 3.18. A many slit diffraction experiment.

Each of the slits gives rise to its own diffraction pattern. All these patterns will give rise to a central maximum on the axis of the system since the condensing lens has the property of bringing rays parallel to the optic axis together at the primary focus, point P in this case. The diffraction patterns of each of the slits are precisely the same. The phase relationships of the patterns arising from the different slits will be different, however, and will tend to cancel unless the position of the diffraction is such that the beams from *all* of the slits are in phase at that point.

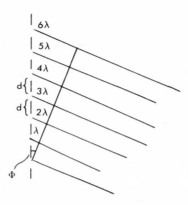

Figure 3.19. The condition leading to a diffraction maximum in a grating experiment.

Figure 3.19 shows the situation for which this is true. At the angle Φ, the contribution of all the slits will be in phase at the viewing screen. Only when such a condition is met will there be a lighted region of the viewing screen. At all other points the screen will be dark.

The distance between the slits is d in Figure 3.19. From the geometry of the figure, we can see that:

$$\sin \Phi = \frac{n\lambda}{d} \qquad (3.8)$$

Equation 3.8 is known as the grating equation.

Gratings are prepared with 10,000 to 30,000 lines (slits) per centimeter. This is a very complicated process which is accomplished by ruling a glass blank with a diamond stylus. The scratches made by the stylus scatter the incident light, while light incident on the regions between the scratches is passed. The regions between the scratches constitute the slits. Original ruled gratings are rarely used. Instead, replicas are made by coating the original with a plastic material and peeling it off after it has set. The plastic is then cemented to another glass blank and this becomes a grating.

Example 3.7. Light with a wavelength of 600 mμ is incident on a diffraction grating with 10,000 lines/cm. How many diffraction orders can be seen?

The maximum value for sin Φ is 1, corresponding to diffraction at 90°. Diffraction through this angle would give an almost undetectable line because of the obliquity factor, which operates in this case. The value 10,000 lines/cm may be expressed as 10^{-4} cm/line $\approx 10^3$ mμ/line. If we put this into equation 8 we get:

$$n = \frac{d \sin \Phi}{\lambda} = \frac{10^3 \times 1}{6 \times 10^2} = 1.66$$

Obviously we cannot see part of an order, and only one 600 mμ maximum will be seen.

The diffraction grating is most commonly used in observing the spectrum of a light beam. It is more valuable than the prism in this type of experiment since, if the grating spacing is known, measuring the angle through which a line is deviated immediately gives the wavelength. With prisms which have nonuniform dispersion, the measurement of wavelength is a much more complicated process. Spectra formed by gratings are said to be rational (dispersion directly proportional to the wavelength) while those formed by prisms are called irrational.

Reflection gratings are also made. With such grating the source and the detector lie on the same side of the grating, and the rulings either scatter or reflect the light. This reflected light behaves in exactly the same fashion as light just passed through a transmission grating. Reflection gratings are often made curved so that the mirror curve of the grating takes the place of the condensing lens.

Example 3.8.   In order to determine the grating spacing for an uncalibrated grating the angular dispersion of the second order of the sodium D line 590 mμ was found to be 70.2°. What was the grating spacing?

From equation 3.8 we see that:

$$d = \frac{n\lambda}{\sin \Phi} = \frac{2 \times 590}{0.941} = 1.25 \times 10^3 \, m\mu$$

and the reciprocal of this is the number of lines per centimeter, which works out to be 8000.

Example 3.9.   Through what angular spread will the first order visible spectrum (400–700 mμ) be dispersed by a grating with 12,500 lines/cm?

A value of 12,500 lines/cm corresponds to 8 $\times$ 10^{-5} cm/line $\approx$ 800 mμ/line. The first order 400 mμ line is spread through an angle of:

$$\sin \Phi = \frac{1 \times 400}{800} = 0.5$$

and $\Phi = 30°$, while the 700 mμ line is dispersed through:

$$\sin \Phi = \frac{1 \times 700}{800} = 0.875$$

and $\Phi = 61°$. Therefore, the first order spectrum is spread through 31°.

Chapter 4

Polarization

The ideas associated with the wave theories of light which we have dealt with so far are independent of the nature of the waves. That is, we have not had any reason to concern ourselves with whether the waves were transverse waves or longitudinal waves. Light waves are transverse waves. We will discuss here a phenomenon which is strictly a property of transverse waves. This phenomenon is called polarization.

Transverse waves are waves in which the displacement of the supporting medium is normal to the direction of propagation. This differs from the disturbance in longitudinal waves where the displacement is along the axis of propagation. Figure 4.1 illustrates the two displacement modes. Transverse waves lack the symmetry about the direction of propagation which is a property of longitudinal waves. This chapter will deal with the question of the direction of the displacement. Figure 4.2 defines a displacement vector D in the x,z-plane. If the vector D remains in the x,z-plane along the entire wave, the wave is said to be *plane polarized*. If D rotates along the length of the wave, the wave is said to be *elliptically*, or *circularly*, *polarized*.

This chapter will deal with the methods by which polarization is achieved and detected. The simplest of these methods is the use of a polarizing material, a material which passes only the light with one plane of polarization. We will then discuss the polarization resulting from reflection and scattering of light. Finally, we will consider circular polarization and elliptical polarization.

Dichroism.   Dichroism is the production of polarized light by use of materials which selectively absorb light having a plane of polarization other than the one specified by the material used in the experiment. The best known and most widely used dichroic material is polaroid sheet, a material of organic origin. Many minerals, as well as organic compounds, exhibit this property.

Polaroid sheet is commonly used in sunglasses, and is a development of the Polaroid Corporation (makers of the Land Camera). It consists of a plastic film in which is suspended large amounts of the compound quinine iodosulphate. This compound

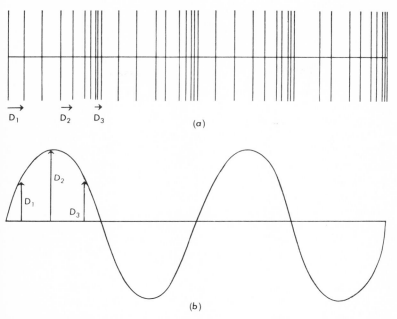

Figure 4.1. The longitudinal (*a*) mode of wave propagation, and the (*b*) transverse mode showing the displacement vector at various points along the wave.

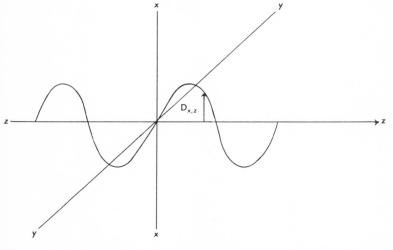

Figure 4.2. A plane polarized wave in the x,z-plane with the displacement vector **D** always in the x,z-plane.

was known to be dichroic early in the nineteenth century. It occurs as fine needle-like crystals, however, and therefore was of
little use by itself. During the manufacture of polaroid sheets,
these crystals are suspended in plastic sheets. The sheets are then
stretched in one direction, giving all the crystals the same orientation. Polaroid sheets can be located between glass plates for
optical uses, thus forming the largest dichroic sheets available.

When light passes through dichroic materials, it appears to be
unaltered except for a small decrease in its intensity. When a
second dichroic sheet is introduced into the light beam, a significant change in the intensity of the transmitted light may occur,
and this intensity can be altered by changing the relative angular
orientation of the sheets. If the sheets have their crystalline axes
parallel as in Figure 4.3, no change in the intensity is noted in the

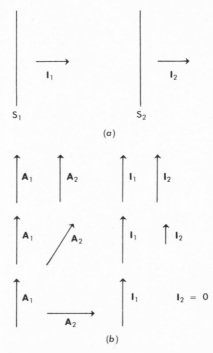

Figure 4.3. (a) Two dichroic sheets with I_1 representing the intensity of the
light after passing sheet 1, and I_2 the intensity after passing sheet 2.
(b) A_1 and A_2 represent the relative orientations of the crystalline
axes of sheets 1 and 2, and I_1 and I_2 the intensities as defined above.
Three relative orientations are shown.

light passing from sheet 1 to sheet 2. If, however, there is an angle between the crystalline axes of the two sheets, a diminution in the transmitted light occurs. If the angle between the axes is 90°, the transmitted intensity falls to zero. This is only consistent with the transverse wave picture of light. No argument based on a longitudinal model of light would explain this result.

In light from an extended source, the production of the light is a random statistical process and is incoherent. Likewise, the relative orientation of the polarization vector (the vector that determines the plane of polarization, $\mathbf{D}_{x,z}$ in Figure 4.2) from various parts of the source is statistically determined and is random. When a beam of this light passes through a dichroic sheet, one plane of polarization is selected by the sheet and only the waves having that polarization can pass. When waves have a polarization at an angle to the crystalline axis of the sheet, the sheet passes only the component of the wave which is parallel to the axis of polarization. This process is illustrated in Figure 4.4.

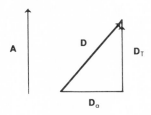

Figure 4.4. **A** is the crystalline axis of the polarizing sheet and **D** is the plane of polarization of the incident light. The relative orientation is shown. **D** is broken into a component $\mathbf{D}_T$ parallel to **A**, which represents the portion of **D** which is transmitted, and $\mathbf{D}_a$ normal to **A**, which is absorbed by the polarizing sheet.

The introduction of a second sheet with its axis at right angles to the first sheet then acts on light which has no component along its plane of polarization and the transmitted intensity falls to zero.

The first sheet in Figure 4.3 is called the *polarizer*. It has the function of establishing a plane of polarization for the beam of light which is under study. Under normal conditions, it is not possible to detect the difference between random light and polarized light. In order to determine the polarization properties of a light beam, a second sheet, S_2, of polarizing material is required. This sheet is called the analyzer. Most dichroic sheets are not quite perfect, but they are adequate for most experimental needs and applications.

As was stated in Chapter 1, light intensity is proportional to the square of the amplitude. A dichroic sheet acts on the amplitude vector and not directly on the intensity. The introduction of a dichroic sheet into a beam of light will result in the intensity being halved since, statistically, one-half of the light has the proper orientation. If the analyzer sheet makes an angle of Θ with the plane of polarization, the amplitude of the final beam is parallel to the axis of the analyzer, as is shown in Figure 4.5.

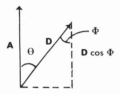

Figure 4.5. The axis **A** of the dichroic sheet and the amplitude **D** of the incident plane polarized beam showing the relative orientation and the component of **D** parallel to **A**.

The intensity, I_2, following passage through S_2 is then $I_1 \cos^2 \Theta$. The introduction of further dichroic sheets into the beam can be handled by treating the sheets serially using the principle established here.

Example 4.1. Three polaroid sheets are placed in a beam of light. The second has its axis at 45° to the first, and the third has its axis at 45° to the second. What fraction of the incident intensity is transmitted.

The first sheet reduces the incident intensity to one-half its original value. The value of $\cos^2 45°$ is $\frac{1}{2}$, so the transmitted intensity is:

$$I_t = \frac{1}{2} \times \frac{1}{2} \times \frac{1}{2} = \frac{1}{8} = 0.125$$

so that 12.5% of the incident intensity is transmitted.

Brewster's Law; Reflective Polarization. One of the most interesting examples of the formation of polarized light was discovered by Malus in 1808. Malus found that light which was reflected from a *nonmetallic* surface was partially polarized by the reflection process itself. The fractional polarization can be altered by changing the angle at which the light is incident on the surface, and there exists a critical angle at which the reflected light is completely plane polarized. Sunglasses made of polaroid sheets are rendered particularly effective near bodies of water by this effect.

Sir David Brewster (1781–1868) studied the effect noted by Malus in some detail. A beam incident on a nonmetallic reflector is split into two beams, the reflected beam and the refracted beam. This is illustrated in Figure 4.6. The reflected beam can be ob-

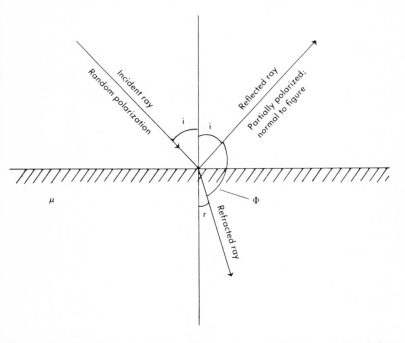

Figure 4.6. The geometry of Brewster's experiment. When the angle labeled Φ in the above figure is 90° the reflected ray is fully plane polarized with the displacement vector normal to the plane of the figure.

served with an analyzer. It is found that the reflected beam is partially polarized with the displacement vector parallel to the surface and normal to the plane of incidence. Brewster noted in his experiments that when the angle between the reflected ray and the refracted ray is 90°, the reflected ray is fully plane polarized, in this case, normal to the figure.

In terms of Figure 4.6, having the reflected ray and the refracted ray at right angles is equivalent to the condition:

$$i + r + 90° = 180° \qquad (4.1)$$

where we have made use of the fact that the angle of incidence, i, is equal to the angle of reflection, r. Solving equation 4.1, we see that

$$i + r = 90° \qquad (4.2)$$

Snell's law [1] gives the relation between i and r as:

$$\sin i = \mu \sin r \qquad (4.3)$$

where we have assumed the external medium to be air. From equation 4.2 we see that:

$$r = 90° - i$$

and therefore:

$$\sin r = \cos i$$

This expression can be substituted into equation 4.3 where we must keep in mind that we are now dealing with the situation where the angle between the reflected and refracted rays is 90°. Making the substitution, we get:

$$\mu = \frac{\sin i}{\cos i} = \tan i \qquad (4.4)$$

This expression relating the angle of incidence to the refractive index for the situation which gives full plane polarization of the reflected ray is known as *Brewster's law*. An example will serve to illustrate its use.

Example 4.2. At what angle above the horizon will the sun be when its light reflected from the surface of a lake is fully plane polarized?

The index of refraction of water is 1.33. Applying Brewster's law (equation 4.4) we get:

$$1.33 = \tan i$$

and i = 53.1°. We know therefore that when the sun is 36.9° above the horizon the reflected light will be plane polarized with

[1] For a discussion of Snell's law see Chapter 2 in *Optics I*. For our present purposes, we shall make the general equation ($n_1 \sin \Phi = n_{II} \sin \Phi'$) more specific, and we let angle Φ be i, angle Φ' be r, and n_{II}/n_1 be μ.

the displacement vector parallel to the surface of the lake. Sun-glasses which make use of the polaroid sheet discussed previously have their crystalline axis vertically oriented, and, therefore, much of the glare produced by reflections from water surfaces is elimi-nated.

A rather old-fashioned method of producing polarized light was based on Brewster's law. A stack of thin glass plates was mounted in a tube as illustrated in Figure 4.7. Glass has a refrac-

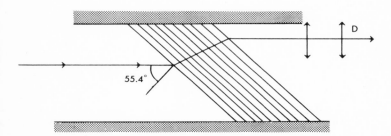

Figure 4.7. The stacked sheet polarizer derived from Brewster's law for glass plates $\mu = 1.5$.

tive index of about 1.5 so that Brewster's angle can be found using equation 4.4. It turns out to be 55.4°. The plates are mounted so that the angle of incidence is 55.4°, and sufficient plates are used so that the many reflections at their interfaces completely remove the component polarized normally to the figure. The transmitted light is plane polarized parallel to the plane of the figure and some displacement vectors are shown.

Closely related to the polarization in reflected light is the polarization in scattered light. There are many orders of light scattering, from scattering by large dust particles in the air to a scattering on the atomic level. It is the dust particles in the air which make the beam of projected light visible in a motion picture theater, and the minute drops of water in the air which make a ray of sunlight visible through storm clouds.

Scattering from the smaller air molecules is quite different from the scattering from dust particles. It is primarily a process involv-ing the wave of light rather than the ray of light as with the larger particles. The wave striking a particle in the atmosphere causes the particle to act as a secondary source for the incident light. However, the secondary source does not scatter the incident light uniformly. The scattering intensity varies inversely as the fourth power of the wavelength. This means that light at 350 mμ is scattered sixteen times more strongly than light at 700 mμ.

Figure 4.8 illustrates the effect of this scattering in the upper atmosphere. Light from the sun which passes through the upper

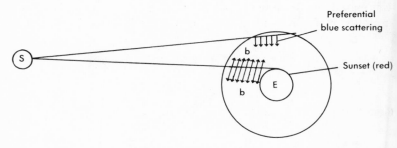

Figure 4.8. Scattering of the sun's light by the upper atmosphere. The arrows indicate the preferential blue scattering giving rise to the blue light of the sky. The sunset beam shows how the blue has been scattered out of the ray giving rise to the reddish appearance of the sun at sunset.

atmosphere is scattered by the constituents of the atmosphere, and, because of the preferential blue scattering, most of the light reaching us from the sky overhead is blue. The atmosphere itself is colorless, but the light reaching us is scattered primarily from the blue end of the spectrum. At sunset, when the direct light from the sun which reaches us travels through a thickness of atmosphere forty-to-fifty times thicker than when the sun is directly overhead, most of the blue has been scattered out of the sun's light and the sun has its characteristic sunset orange appearance.

The old sailor's saying, "red sky in the morning, sailor take warning; red sky at night, sailors delight," has a basis in fact since only when a high pressure region is to the west of you will you see a red sunset. The high pressure is needed to make the path of the sunlight in the atmosphere long enough to eliminate the blue component. High pressure to the west usually means good weather approaching. High pressure to the east would mean a red sunrise and probably bad weather.

Examination of the light scattered by small particles such as sky light shows that the scattered light is plane polarized. The

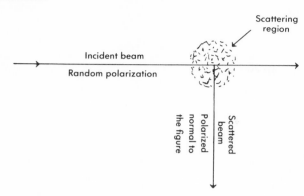

Figure 4.9. Polarization by scattering. The scattered beam is polarized with the plane of polarization normal to the figure.

polarization takes place in the plane normal to the direction of the primary beam. Figure 4.9 shows the geometry of this process which can be easily observed in sky light with an analyzer.

Double Refraction. We have previously considered refraction effects in *Optics I* of this series, but in dealing with refraction we have only considered isotropic materials such as water and glass. *Isotropic materials* have uniform physical properties in all directions within the body. In Figure 4.10, we see the refracted image of a line through a piece of glass and through a piece of calcite crystal which has been carefully cut and polished. The glass forms a single refracted image, as we have come to expect, but the calcite crystal produces two images of the line. This effect is known as *double refraction*, and is found only in certain anisotropic materials.

A common experiment done with doubly refracting materials is illustrated in Figure 4.11. A spot is marked on a sheet placed

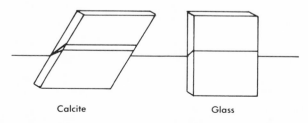

Calcite   Glass

Figure 4.10. Double refraction by a calcite crystal. The glass plate on the right illustrates the image formed by an isotropic material.

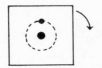

Figure 4.11. A calcite crystal resting on a sheet marked with a dot. The view is
normal so that one would expect the image to lie on the viewing
axis. Actually one image is on the viewing axis but a second has
been formed above the first. As the crystal is rotated this second
image rotates around the axis of the first as is shown.

below the calcite crystal. If the crystal is viewed from above and
is rotated, the anomalous second spot rotates about the primary
image formed on the axis immediately above the spot. If the two
images are viewed through an analyzing polarizer, they are found
to be plane polarized with the plane of one being at 90° to the
plane of the other. Clearly, double refraction is associated with
some polarization effect.

Figure 4.12 illustrates schematically what occurs when a beam
of light with random polarization is incident on a calcite plate.
The beam is split into two components with different polarization
planes. The ordinary ray, labeled O, is polarized in the plane of
the figure, while the extraordinary ray labeled E is polarized

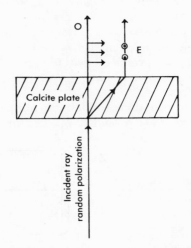

Figure 4.12. The ordinary ray O and the extraordinary ray E in a calcite crystal
showing the polarization of O parallel to the figure and E normal
to the figure.

normally to the figure. The anistropy which exists in calcite is a variation of index of refraction (and therefore of the velocity of light) in various directions within the crystal.

Figure 4.13 shows the variation of velocity of propagation of the O-wave and the E-wave as a function of direction within the crystal. The optic axis is defined as that direction in the crystal along which the O-wave and E-wave have the same velocity. Since refractive properties depend upon the index of refraction (velocity of propagation), it is not surprising that two distinct velocities for the two different modes of polarization would lead to the

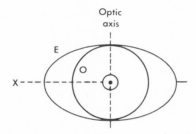

Figure 4.13. The velocity of propagation as a function of direction in a calcite crystal. In the direction in which the velocities of the O wave and the E wave are the same we have drawn a dotted line. This direction in the crystal is known as the optic axis.

formation of two refractive images. Knowledge of the two velocities of propagation can be combined with Snell's law to predict the position of the two images, if such cannot be measured.

Various materials which appear to be isotropic become doubly refracting when subjected to strong electrical fields (*electro-optic effect*) or strong magnetic fields (*Cotton-Mouton effect*). These effects are used to switch beams of light with established polarizations in various commercial applications.

Circular Polarization. Suppose that we cut a slice of calcite so that the broad face of the slice is parallel to the optic axis and the face is cut normally to the axis OX of Figure 4.13. The resulting slab of material will propagate the E-wave much more quickly than the O-wave, and this will result in a relative phase shift between the E and O-waves. The geometry of the slice is shown in Figure 4.14.

The polarization vectors of the O-wave and the E-wave are also shown in Figure 4.14. If plane polarized light falls on the calcite

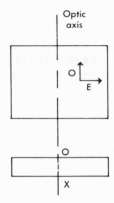

Figure 4.14. The geometry of a half-wave and a quarter-wave plate.

with its polarization vector parallel to the O vector or the P vector, there is no change in the transmitted light. We want to consider what happens when the plane polarized light falls on the calcite crystal at an angle of 45° with each of the vectors E and O.

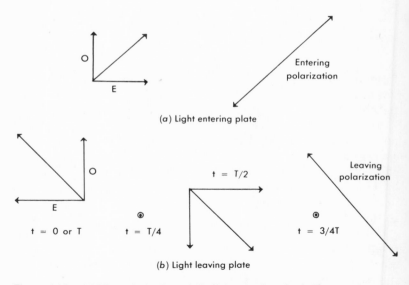

Figure 4.15. (*a*) The polarization of the light entering the half-wave plate and its E and O components. (*b*) The light leaving the half-wave plate at four intervals during one period of the wave. The net polarization is then at right angles to the incident light.

The plane polarized incident light will split its intensity into the components parallel to E and parallel to O, as in Figure 4.15.

The velocity of propagation of the E polarized component will be greater than the velocity of the O component. This is clear from the relative velocities shown in Figure 4.13. The nature of the polarization which one observes in the light coming from

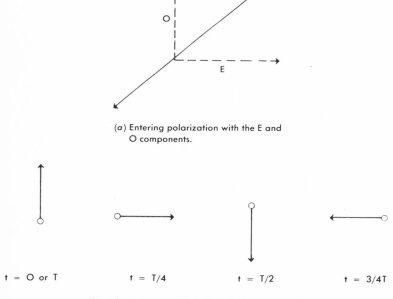

(a) Entering polarization with the E and
    O components.

t = O or T   t = T/4   t = T/2   t = 3/4T

(b) Polarization at various times during one period
    of light leaving the quarter-wave plate.

(c) Locus of the exiting polarization vector.

Figure 4.16. The effect of a quarter-wave plate on plane polarized light and right circular light.

the calcite slice will depend upon the thickness of the slice. Zero thickness results in no change in the polarization vector. This zero shift will also be seen if the plate is so thick that the E-wave is phase shifted by 360° relative to the O-wave.

If the thickness of the calcite plate is such that the E-wave is shifted 180° relative to the O-wave, the plate is known as a half-wave plate. The result of this can be seen in Figure 4.15. The figure shows one full cycle of the light emerging from the crystal. The result of inserting a half-wave can be seen to be the rotation of the plane of the incident plane polarized light through an angle of 90°.

The most interesting case is seen when the plate produces a 90° phase shift in the E-wave relative to the O-wave. Such a plate is known as a quarter-wave plate. Figure 4.16 shows the entering polarization and the polarization of the emerging wave at four times during one period. The result of placing this plate in a beam of plane polarized light is to cause the locus of the head of the polarization vector to have geometry shown in 4.16c. Such light is called circularly polarized light. Elliptically polarized light can be produced using a quarter-wave plate if the polarization vector of the incident light makes an angle of other than 45° with the optics axis.

If the light emerging from the plate has a polarization vector which appears to rotate in a clockwise direction, it is said to be right circular (or elliptical) light; counter clockwise rotation is called left circular light.

Polarimetry. Certain materials, of which sugar is the most common example, when placed in solution cause the plane of polarization of plane polarized light to rotate. The rotation depends on the concentration of the solution and the length of the path through the solution. Materials which cause such rotation are called *optically active* and are classed *dextro-rotary* or *levo-rotary* if they rotate the plane of the polarized light to the right or to the left, respectively. The purity of sugar, which is the largest tonnage pure chemical, is analyzed by studying the specific rotation (rotation per unit concentration) of each lot sold.

Chapter 5
Quantum Optics

In *Optics I* of this series we considered the ray model for the propagation of light. In this volume we have discussed the wave model. This chapter will be devoted to the particle model. During his studies of optics, Newton held that light was a stream of small particles. The interference and diffraction studies made during the eighteenth and nineteenth centuries were used as the basis for the argument that light was essentially wavelike. The arguments for wavelike propagation were accepted even more fully with the rise of the electromagnetic theory due to the work of Maxwell. The view accepted at present by most scientists is that light is neither a particle nor a wave, but rather some combination of the two. Our ability to explain interference and diffraction effects on the basis of the wave model suggests that wave properties must be present, while the experiments that we will discuss in this chapter certainly suggest a particle nature. Generally, we use the model which has the properties required by the experiment.

Planck's Hypothesis. One of the most outstanding physical problems which had not been solved by the end of the nineteenth century was the question of black-body radiation. A *black body* is an object which is a perfect absorber of radiation. That is, all radiation which falls on the black body is absorbed. Such a body is also a perfect radiator. The rate of radiation by a black body is proportional to the fourth power of the absolute temperature, T. The power radiated obeys the *Stefan-Boltzmann law*:

$$P_{black} = \sigma T^4 \tag{5.1}$$

where σ is a proportionality constant and has the value 5.670×10^{-8} watts/m^2 − (K°)^4.

Example 5.1. How much power is radiated by a black-body sphere 1 m in radius if its temperature is 600°C?

A temperature of 600°C is equivalent to 873°K. When we substitute this into equation 5.1, we find that:

$$P = 5.670 \times 10^{-8} \times (873)^4 \text{ watts/m}^2 = 3.29 \times 10^4 \text{ watts/m}^2$$

The sphere has a radius of 1 m, so that it has an area of 4π m^2 and the total energy radiated is 41.3×10^4 watts.

While the Stefan-Boltzmann equation gives the total energy radiated by a black body, it gives no information about the spectrum of the emitted radiation. In order to discuss the spectrum of emitted radiation, it is necessary to introduce a quantity known as the spectral radiance, the energy emitted per unit wavelength at some wavelength λ. The spectral radiance is measured in watts/m^2 per millimicron. Figure 5.1 shows the measured spectral radiance for three different temperatures. These curves

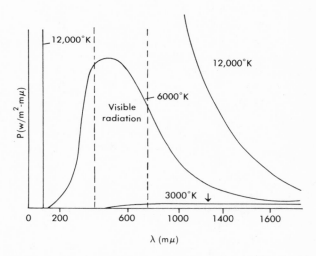

Figure 5.1. The black body radiation curves for three different temperatures. The 6000°K curve is characteristic of the radiation for the sun.

have several important features. They never cross, and the curves for higher temperatures lie above the curves for lower temperatures. The maximum shifts toward shorter wavelengths as the temperature increases. This shift results in the well-known characteristic change in the appearance of a body as the temperature is increased. The body will first glow red, then white, and finally blue as the temperature increases and the wavelength of the radiation maximum increases. As a rule, the blue is only seen in some very hot stars. The 6000°K curve is characteristic of our sun, and can be used to define white light.

During the last decade of the nineteenth century, a great deal of effort was expended by physicists in order to derive an analytic expression for the black body curve from electromagnetic theory. Their efforts were unsuccessful until, in 1900, the German physicist Max Planck (1858–1947) made his famous quantum hypothesis. This hypothesis assumed that light was emitted and absorbed in discrete bundles called *quanta* or, more commonly, *photons*. This idea is completely contrary to the wave picture we have previously discussed. A wave theory would make emission and absorption a continuous process rather than a quantum process.

Planck's quantum hypothesis may be stated:

> *Radiation cannot be emitted or absorbed in an arbitrary amount, but is always emitted or absorbed in discrete quantities known as quanta. For radiation of frequency v, each quantum has an energy of hv, where h is a universal constant.*

It follows immediately from Planck's hypothesis that an integral number of quanta must exist in any radiation field, since nonintegral numbers of quanta are not allowed in Planck's theory.

Using this hypothesis, Planck was able to derive an analytic expression for the black body radiation curves which fit them to within the limit of experimental error. Planck's constant, as the quantity h has been named, has a value of 6.6256×10^{-34} joule-sec. In a later section, we will see how this constant is determined. The units of h are energy-time and are known as action units.

Example 5.2. How much energy is contained in each quanta of radiation at 560 mμ? How many quanta are emitted per second from a source radiating at 1 watt?

Planck's hypothesis can be written:

$$E = h\nu \qquad (5.2)$$

The problem gives the wavelength rather than the frequency, ν, so we will modify equation 5.2 by substituting c/λ for the frequency, which then makes equation 5.2 read:

$$E = \frac{hc}{\lambda} = \frac{6.63 \times 10^{-34} \times 3 \times 10^8}{560 \times 10^{-9}} = 3.55 \times 10^{-19} \text{ joules}$$

Each photon of wavelength 560 mμ therefore carries 3.55×10^{-19} joule energy. Since one watt is one joule per second, the number of photons is equal to:

$$\frac{1 \text{ joule/sec}}{3.55 \times 10^{-19} \text{ joule/photon}} = 2.82 \times 10^{18} \text{ photons/sec}$$

At maximum sensitivity, the eye can detect about 250 photons per second falling on it (if the conditions are optimized). Very sensitive photodetectors can detect single photons.

The Photoelectric Effect.   The photoelectric effect experiments were analyzed by Einstein in 1905 using Planck's hypothesis. It was the results of this analysis which gave substance to the idea of quanta of light and revived the particle picture of light.

It is well known that hot metals emit electrons. It is this thermionic emission of electrons which allows one to construct vacuum tubes. Electrons can also be emitted by certain metals when illuminated by light. Electrons are emitted when light of sufficiently high frequency (short wavelength) falls on a metallic surface which has been properly prepared.

Figure 5.2 shows the apparatus used in studying the effect. The metal surface, A in the figure, is illuminated by light of energy $h\nu$. Electrons are emitted from the metallic surface and are attracted to the cathode, C, which is maintained at a voltage, E, above the level of the photosurface. Electrons reaching the cathode flow to the battery through the galvanometer and are recorded by the galvanometer reading.

The experiment produces a number of results which must be explained. No electrons are observed until the frequency of the photons reaches some threshold value which is characteristic of the metal being used in the experiment. Electrons appear as soon as the illumination begins if the frequency of the light quanta is above the threshold frequency. Finally, the energy of the electrons emitted increases as the frequency of the illumination increases. This does not exceed some fixed value for any frequency, however.

If we were to assume that light arrived at the surface of the metal as waves, then, regardless of the frequency of the light, after a sufficient amount of time enough energy should arrive at the surface to cause an electron to be ejected. One sees no electrons until the threshold has been reached, however, and then electrons immediately appear at the cathode. A wave picture would require a finite period of time until enough energy could be carried in by the wave to the electron to give it an escape energy.

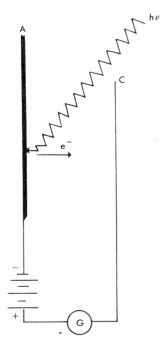

Figure 5.2. Apparatus for the study of the photoelectric effect. Light quanta striking the photosurface A cause electrons to be ejected and these electrons are attracted to the cathode C by the voltage set by the battery. The flow of current is observed in the galvanometer.

Finally, based on a wave picture, one would expect that as the intensity of illumination increased, so would the energy of the escaping electrons. This is, of course, not what is observed.

Electrons in metals have an energy less than zero. This means that it is necessary to do a certain amount of work on the electron to remove it from the metal. This amount of work is dependent on the crystalline structure of the metal and on the surface condition. It is therefore not surprising that the amount of energy required to remove the electrons from each metal is different, but that this energy remains the same in different experiments on the same metal.

In order to establish the maximum energy of the escaping electrons an apparatus such as is illustrated in Figure 5.3 is used. The retarding voltage, V, is slowly increased until no electrons

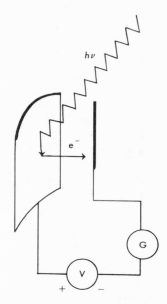

Figure 5.3. Apparatus for measuring the maximum kinetic energy of the photoelectrons emitted in the photoelectric effect. The voltage is set to
repel the photoelectrons and is varied until it just shuts off the
electron current.

reach the cathode. By carefully plotting the maximum energy of
the photoelectrons as a function of frequency, curves such as
those illustrated in Figure 5.4 are obtained for each metal.

In Figure 5.4, the curves for the frequency versus the maximum energy, KE_{max}, turn out to be straight lines which are
parallel for different metals. Einstein fitted these curves to an

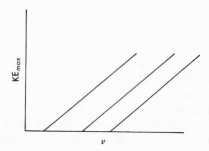

Figure 5.4. KE_{max} vs. frequency for three different metals.

expression:

$$KE_{max} = h\nu - \Phi \qquad (5.3)$$

where Φ is a constant energy known as the *work function* characteristic of the metal being studied. The constant of proportionality which is written above as h of course turns out to be Planck's constant. We can therefore conclude that a particle picture is consistent with this kind of experiment in optics.

One of the significant results of Einstein's analysis is that it gives a method for the measurement of Planck's constant. The slope of the lines drawn in Figure 5.4 from experimental data is in each case h. One can therefore evaluate Planck's constant to several significant figures by measuring the photo response of several metals. This is one of the best ways of making this measurement, since the possibility of error is cut down by the use of different metals in the measurement.

Example 5.3. The work function for a tungsten surface may be determined from the photo threshold. This threshold is found to be 273 mμ. What is the work function? If the surface is illuminated by light of wavelength of 150 mμ, what is the maximum kinetic energy of the ejected photoelectrons?

At the threshold, the kinetic energy of the ejected photoelectrons is zero. Therefore, from equation 5.3 we see that:

$$0 = h\nu - \Phi$$

$$\Phi = \frac{hc}{\lambda} = \frac{6.63 \times 10^{-34} \times 3 \times 10^{8}}{273 \times 10^{-9}}$$

$$\Phi = 7.2 \times 10^{-19} \, J.$$

To find the KE_{max}, we substitute the work function Φ and the wavelength as given in the problem into equation 5.3 and get:

$$KE_{max} = \frac{6.63 \times 10^{-34} \times 3 \times 10^{8}}{150 \times 10^{-9}} - 7.2 \times 10^{-19}$$

$$KE_{max} = 6.06 \times 10^{-19} \, joule$$

Emission and Absorption of Radiation. Earlier in this chapter we considered one kind of optical emission; that resulting from the temperature of a system. We will now discuss another form of emission: namely the emission of radiation by individual atoms. The subject of atomic emission is treated most fully by a branch of physics known as quantum theory. Quantum theory is beyond the scope of this text and so we will discuss atomic emission in a rather qualitative fashion.

Individual atoms do not radiate with a radiation continuum as do black bodies. Rather, atoms tend to radiate only certain discrete frequencies or wavelengths. Figure 5.5 shows the most intense spectral lines which appear when the elements sodium, cadmium, and mercury are excited and forced to emit. Optical

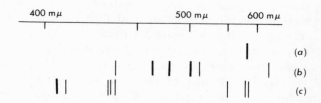

Figure 5.5. The strong lines in the visible emission spectra of (a) sodium, (b) cadmium, and (c) mercury.

emission occurs when an electron associated with some atom is excited to a higher, unoccupied energy level and then falls to a lower energy level. The energy lost by the electron appears as a single photon. Such transitions are called radiative transitions. Figure 5.6 illustrates the process.

Other decay processes do occur. These processes, which generally involve inelastic collisions between atoms, have the excess energy carried away by the colliding atom and are called nonradiative transitions.

Einstein studied the radiative transition probability of an ex-

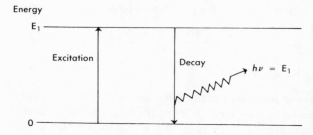

Figure 5.6. The radiative decay process. The atom is excited from its ground state (E = 0) to an excited state at E_1. The atom then decays through the emission of a photon with $E_1 = h\nu$.

cited atom and determined that two processes can occur. Atoms may decay by either a spontaneous process which takes place statistically, or by stimulated decay which is induced whenever other radiation of the same frequency is present. The stimulated decay emission is proportional to the spontaneous decay probability, but also depends on the amount of radiation of the same frequency which is present. Stimulated decay can lead to very intense radiation from a collection of excited atoms. Einstein showed that the probability of excitation and decay are the same for any atom, but only one excitation process can occur while the two decay processes discussed above exist.

One might conclude from the above that for every decay which takes place in some system an excitation also takes place. If this were so, we would not be able to observe absorption of radiation. Actually, the difference between absorption and decay which we see arises because, in any collection of atoms, generally more atoms are in the ground state than are in excited states. This population difference in states exists whenever a system is in thermal equilibrium. Because there is a difference in the number of atoms undergoing either absorption or decay, we are able to observe these processes.

The equilibrium population at any temperature T (absolute) is governed by Boltzmann statistics. If N_1 is the number of atoms per unit volume in the excited state E_1 of Figure 5.6 and N_0 is the number of atoms in the ground state, then the populations are related by the expression:

$$\frac{N_1}{N_0} = e^{-\Delta E/kT} \tag{5.4}$$

where k is the Boltzmann constant 1.38×10^{-23} joules/°K and ΔE in this case is: $E_1 - 0 = E_1$. As a result of the fact that there are always more atoms in the ground state than in the excited state at thermal equilibrium, and the fact that the probability of absorption and decay are equal, a system in thermal equilibrium will always show a net absorption.

Example 5.4. If E_1 in Figure 5.6 is 1.6×10^{-19} joules and the temperature is 300°K (room temperature), what fraction of the atoms will be in an excited state?

We can begin by evaluating the exponent in equation 5.4.

$$\frac{\Delta E}{kT} = \frac{1.6 \times 10^{-19} \text{ joules}}{1.38 \times 10^{-23} \text{ joules/°K} \times 300°K} = 38.6$$

The ratio N_1/N_0 is the fraction of atoms in the excited state. Since we know the exponent form equation 5.4, we can substitute in 5.4 and get:

$$\frac{N_1}{N_0} = e^{-38.6} = 1.6 \times 10^{-17}$$

Roughly, only one atom in 10^{17} will be in an excited state under the prescribed conditions.

The Laser. We have seen that it is possible to have atomic absorption take place in thermal equilibrium. Suppose that we were to seek a nonequilibrium situation in which the number of atoms in excited states exceeds the number in the ground state. This situation leads to atomic emission spectra. Figure 5.5 illustrates some typical emission spectra. If the rate of emission of the inverted population can be controlled, it should be possible to amplify light. This was achieved as early as 1952 by Charles H. Townes of Columbia University, working in the microwave region. Townes was able to produce a population of ammonia molecules in an inverted ($N_1 > N_0$) population. This device was called a maser, which is an acronym for *m*icrowave *a*mplification by *s*timulated *e*mission of *r*adiation. Light amplifiers using an inverted population are called lasers—*l*ight *a*mplification through *s*timulated *e*mission of *r*adiation.

There are now a large number of experimental techniques for laser amplification. Most of these systems depend upon the existence of a third state, as illustrated in Figure 5.7. The laser system at thermal equilibrium has most of its atoms in the ground

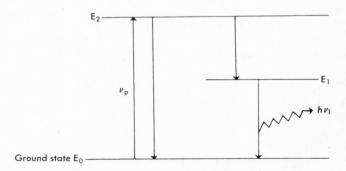

Figure 5.7. An energy level diagram for a three level laser system. The pumping frequency is ν_p and the system radiates at frequency ν_1. The transition E_2 to E_1 is often a radiationless transition.

state, E_0. The system is excited from E_0 to E_2 by a technique called optical pumping. Intense radiation at frequency ν_p "pumps" the atoms with ground state energy into the excited state, E_2. The radiation is either supplied continuously or in some systems by a discharge flash. This process, which is called *optical pumping*, is applied in many of the presently used laser systems. Optical pumping (a process which is the inverse of photon emission) is the Einstein absorption process.

The excited atoms in state E_2 can fall back to the ground state or they can fall into state E_1. E_1 is a metastable state, that is, a state whose lifetime is long compared to the lifetime of E_2. Many metastable states are known to exist in various systems and each is potentially useful for laser design. Since the probability of spontaneous decay of E_1 to the ground state is small, atoms will accumulate in E_1 until the population of atoms in E_1 exceeds the population in the ground state level. This produces the population inversion which is requisite for light amplification.

If, after having produced the inverted population, a beam at frequency ν_1 passes through the assembly of excited atoms, the beam will be amplified. An interesting feature of beams formed by stimulated emission is that they are coherent. The process of stimulated emission occurs with a definite phase relationship between the stimulating wave and the wave emitted by the atom undergoing decay. This is different from spontaneous decay, which is a statistical process and which therefore shows no coherence between decays of different atoms. In addition, the direction of the emitted photon is the same as the direction of the stimulating radiation. The coherence is so strong that two different lasers using the same decay process to produce light may interfere with one another. This is the only known case of interference being seen between two different independent sources.

Figure 5.8 is a simple diagram of a typical laser system. When the optical pumping lamp is discharged, we get an inverted population in state E_1. Figure 5.8b illustrates the method used to make best use of the effect of the flash. The pump and the laser rod are set at the foci of an elliptical mirror which extends the entire length of the system. Elliptical mirrors have the property that all rays passing through one focus are reflected to the other focus so that all the pump light falls on the laser tube. The two mirrors at the ends of the optical system are set so that all light parallel to the axis of the tube remains parallel to that axis after reflection. The mirrors are set apart by a distance equal to some integral

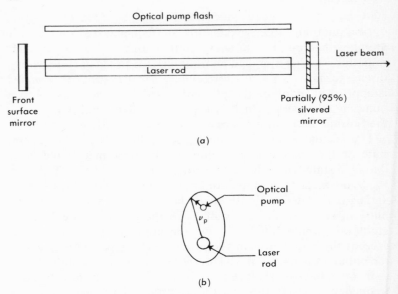

Figure 5.8. A simple laser system. (*a*) The details of the pumping arrangement showing the mirrors used to create standing optical waves. (*b*) The housing of the pump and laser rod showing the elliptical mirror used to focus the entire pump discharge on the laser rod.

multiple of wavelength so that optical standing waves are created in the system. These standing waves occur only because of the coherence of radiation produced by stimulated emission.

One of the mirrors used in the system is only partially silvered so that a fraction (about 5% in this case) of the laser beam can escape the system and be used for various purposes. The other mirror is full-surfaced and reflects all incident radiation.

An operating cycle would occur in the following sequence. The pump is discharged, producing an inverted population in the laser rod. One of the excited states, E_1, spontaneously emits a photon which is along a line parallel to the tube axis. (Many of the spontaneously emitted photons are, of course, lost because they are not emitted along the proper axis.) Once a photon with the proper direction has started, standing waves are set up in the laser rod, and an intense coherent beam is built up. This beam contains enormous intensity and the emergent beam is in a very narrow pencil of light. Many of the applications of the laser beam make use of the fact that the beam does not diverge. An experiment has been performed in which a laser beam was bounced off the moon. The beam only had a three mile diameter on the moon at a distance of 250,000 miles!

The laser described above is a pulsed laser. That is, it does not produce a continuous beam of light, but rather a beam which comes out in short pulses. Because of the enormous energy involved in this process, the laser must be given a short period to dissipate the excess energy it has acquired from heating due to the pulse. Other lasers operate in a continuous mode by using a method of pumping which is not derived from a flash. These continuous lasers radiate much less energy than the pulsed lasers as a rule and are the kind used for most laser demonstrations.

Appendix

Fraunhofer Lines of the Spectrum with their Color

Fraunhofer Line	Wavelength (mμ)	Color
A	759.38	red
B	686.72	red
C	656.28	red
D_1	589.59	yellow
D_2	589.00	yellow
F	486.13	blue
G'	434.05	violet
h	410.19	violet
H	396.85	ultraviolet
K	393.37	ultraviolet

In general:

Color	Wavelength (mμ)
violet	less than 450
blue	450–500
green	500–570
yellow	570–590
orange	590–610
red	longer than 610

Problems

Chapter 1.

1. A lighthouse keeper notes that the seas passing his light-house are 12 ft apart, and that they pass the lighthouse at a rate of 5 waves per second. What is the velocity of wave propagation?

   Ans. 60 ft/sec.

2. The wavelength of sound waves issuing from a particular source is 1 ft, and the velocity of propagation of sound is 1000 ft/sec. What is the period of the sound wave?

   Ans. 0.001 sec.

3. Two waves are observed, one of which has an amplitude of 1.5 ft, and the second an amplitude of 2.2 ft. How much energy is there in the second wave compared to the first?

   Ans. $E_2/E_1 = 2.15$.

4. Derive the equation for the wave profile of a wave traveling the negative x-direction using the same reasoning we used to arrive at equation 1.6.

5. Express equation 1.7 in terms of the wavelength and the period.

   $$Ans. \quad f(x,t) = A \cos 2\pi \left(\frac{x}{\lambda} \pm \frac{t}{T} \right).$$

6. Two sine waves travel together along the positive x-axis of some system. The first amplitude is two units and the second is one unit. The second lags behind the first by $\pi/3$ radians. These combine to form a single sine wave. Find the new amplitude and phase.

   Ans. $\delta = -19.2°; A = 2.65$.

7. Roughly draw out the curves to check the answer to problem 6.

Chapter 2.

1. Young's experiment is performed with two slits fixed on 0.3 mm separation and with the viewing screen at a distance of 0.5 m. If the source has a wavelength of 600 mμ, where does the first bright maximum lie relative to the central maximum?

    Ans. 1 mm.

2. Young's experiment is performed with slits 1.0 mm apart and with the viewing screen at a distance of 1 m. The first bright fringe is 0.5 mm from the central maximum. What is the wavelength of the source?

    Ans. 500 mμ.

3. Suppose Young's experiment is performed in such a way that there is a π phase difference between the light emitted at each of the two slits. Will the central fringe be bright or dark? Write an expression for the distance from the central fringe to the first dark fringe.

    Ans. dark; $y = 1 \times \dfrac{\lambda}{d}$

4. The Lloyd's mirror experiment can be set up with a source of 589 mμ, a plane mirror 0.1 mm below the level of the source, and a screen 1 m from the source. How far above the level of the mirror does the first bright fringe appear? Remember that the mirror will introduce a phase shift in the reflected light.

    Ans. 1.48 mm.

5. A Michelson interferometer is used to measure the wavelength of a particular spectral line. When the moveable mirror has been driven far enough for the observer to have counted 1000 fringes passing the telescope pointer, the micrometer drive indicates that the mirror has shifted 0.156 mm. What is the wavelength of the spectral line?

    Ans. 624 mμ.

6. How many fringes will pass the pointer of the telescope in a Michelson interferometer if the mirror is driven 0.1 mm and a spectral line of wavelength 445 mμ is being observed?

    Ans. 890.

7. A plano-convex lens with a radius of curvature of 200 cm is placed on an optical flat. The lens is then illuminated from above by a source which gives light of wavelength 540 mμ. Where does the fiftieth dark fringe occur? (Remember the central fringe is dark!)

Ans. 0.735 cm from the center.

8. What is the minimum thickness of a soap film of a refractive index 1.34 which gives constructive interference of light of wavelength 589 mμ when the light is normally incident on the film?

Ans. 110 mμ.

9. A Fabry-Perot interferometer is designed to give maximum transmission at 400 mμ. What is the separation of the plates?

Ans. 200 mμ.

10. A lens coating of index 1.45 is used to coat a lens of index 1.55 nonreflecting at 560 mμ. The coating is made $\frac{3}{4}\lambda$ thick instead of the usual $\frac{1}{4}\lambda$ thick. At what wavelength in the visible spectrum will the reflection be a maximum? Remember that λ is modified by the coating material.

Ans. 420 mμ.

Chapter 3.

1. A slit 0.25 mm in width is used to form a Fraunhofer diffraction pattern with a lens of focal length 25 cm. Where is the third diffraction minimum found? $\lambda = 500$ mμ.

Ans. 1.5 mm from the central maximum.

2. How wide is the central maximum of a Fraunhofer diffraction pattern formed with a slit 0.0014 cm wide using a lens of focal length 100 cm? Solve for 700 mμ light.

Ans. 51.7 cm.

3. Parallel light of wavelength 600 mμ is passed through a slit 0.36 mm wide and is focused on a screen by a lens. The first maximum is 2.0 mm from the central maximum. What is the focal length of the lens?

Ans. 80 cm.

4. Monochromatic light of wavelength 400 mμ falls on a circular opening in a plate. Along the axis of the opening it is found that, as an observing screen is brought toward the opening, the first observable minimum occurs at 160 cm. Find the diameter of the aperture.

    Ans. 2 × 0.113 cm.

5. The radius of a hole in an opaque screen is 1.28 mm. At what distance from this screen will an observer see the most intense maximum for 550 mμ light.

    Ans. 3 m.

6. What is the limit of angular resolution for the Palomar telescope with a 200 in. reflecting mirror? (Use 560 mμ.)

    Ans. 1.37 × 10^{-7} radians or 0° 0′ 0.0283″.

7. The moon is 250,000 miles from the earth. How far apart on the moon must two objects be in order that the Palomar telescope will resolve them?

    Ans. 176 ft.

8. A diffraction grating has 4000 lines/cm. What is the angular deviation of the second order sodium (590 mμ) line?

    Ans. 28.1°.

9. In order to calibrate a diffraction grating, light of 515 mμ is used with the grating. The first order spectrum is found at 18°. What is the grating constant?

    Ans. 6000 lines/cm.

10. Light of wavelength 600 mμ falls on a diffraction grating with 5000 lines/cm. Where does the third order maximum appear?

    Ans. 64.2°.

Chapter 4.

1. Given two polarizers at angles of Φ to each other, what fraction of random incident light will be transmitted?

    Ans. $\frac{1}{2} \cos^2 \Phi$.

2. Three polarizing plates are stacked with the first and third parallel to each other and the second at 45°. What is the

intensity of the transmitted beam if random light is incident? What change will be seen in the intensity of the transmitted light if the first and third plates are at right angles and the second is at 45°?

Ans. 1/8; none.

3. Given six stacked polarizers, each at 60° to the previous one, what fraction of the incident random light will be transmitted?

Ans. $\frac{1}{2048}$.

4. What is the Brewster angle for quartz of index 1.533?

Ans. 56.9°.

5. Can the Brewster angle be less than 45°?

Ans. no; 45° corresponds to an index of refraction of 1, which is free space and the minimum index.

Chapter 5.

1. Plot the power radiated by a black body as a function of T and as a function of T^4.

2. How much power is radiated by a black body at room temperature (25°C) if the black body is a sphere 1 cm in radius?

Ans. 0.564 watts.

3. How much energy is carried by a photon of microwave radiation $\lambda = 10$ cm; by infrared radiation $\lambda = 1\mu$; by an X-ray photon $\lambda = 0.1$ mμ?

Ans. 1.99×10^{-24} joule; 1.99×10^{-18} joule; 1.99×10^{-15} joule.

4. A source radiates 1 milliwatt at 600 mμ. How many photons per sec are radiated?

Ans. 3.01×10^{15} photons.

5. Light of wavelength 300 mμ is used to determine the work function of zinc. The measured energy maximum of the emitted electrons is 0.8×10^{-19} joule. What is the energy of each photon, and what is the work function?

Ans. 6.63×10^{-19} joule; 5.83×10^{-19} joule.

6. The threshold for tungsten is 273 mμ. What is the maximum
    kinetic energy of electrons generated by 200 mμ ultraviolet
    radiation?

    Ans.   2.67 $\times$ 10^{-19} joule.

7. A gas laser emits photons at 623.8 mμ. What is the energy of
    each photon? What is the separation of the states producing
    this radiation? What is the ratio of the population of the
    excited state to that of the ground state at 25°C?

    Ans.   3.14 $\times$ 10^{-19} joule; 3.14 $\times$ 10^{-19} joule;
             2.5 $\times$ 10^{-34}.

Index